Citizen's Right to the Digital City

Marcus Foth · Martin Brynskov · Timo Ojala
Editors

Citizen's Right to the Digital City

Urban Interfaces, Activism, and Placemaking

Epilogue by Saskia Sassen

Springer

Editors
Marcus Foth
Urban Informatics Research Lab
Queensland University of Technology
Brisbane, QLD
Australia

Timo Ojala
University of Oulu
Oulu
Finland

Martin Brynskov
Participatory IT Centre
Aarhus University
Aarhus
Denmark

ISBN 978-981-13-5727-5 ISBN 978-981-287-919-6 (eBook)
DOI 10.1007/978-981-287-919-6

Springer Singapore Heidelberg New York Dordrecht London

Softcover re-print of the Hardcover 1st edition 2015

Printed on acid-free paper

Springer Science+Business Media Singapore Pte Ltd. is part of Springer Science+Business Media (www.springer.com)

Preface

In 2006, TIME magazine published a picture of a computer on its cover as the "Person of the Year" with the words:

> You. Yes, you. You control the Information Age. Welcome to your world.

The magazine's selection highlights the profound shift in the way that the World Wide Web has advanced to allow an increasing number of everyday people to not only access information, but also contribute and participate in their own right. This trend has been popularised as "Web 2.0" or "the social media revolution"—Jenkins (2006) coined the term, "participatory culture", heralding a new era of "open government". The perhaps utopian prospect of a democratisation of knowledge through the ability for everyone to become a creator, publisher, remixer, recommender, sharer, and referrer has led to an exponential growth in content: Every 60 s on Facebook, 510,000 comments are posted, 293,000 statuses are updated, and 136,000 photographs are uploaded (thesocialskinny.com). However, quantity of content does not imply quality and thus, with more sources of content, spread via more digital media channels, to more people, Web users started to face the problem of information overload.

Large corporations in the digital economy such as Facebook, Google, Twitter, and Amazon have deployed sophisticated filters and recommendation systems designed to help us navigate the otherwise bloated social mediascape. The content displayed on Facebook's newsfeed is selected based on a user's profile, their location, interests, habits, and online transactions—what they post, share, recommend, and "like". The popularity of social media stems from its power to create personalised spaces and walled gardens, which are tailored to individual preferences and favour content relevant to each user. Algorithms proprietary to each social media site determine what is deemed relevant: with the absence of a journalistic or editorial code of ethics, these machine learning algorithms determine the make-up of the Facebook newsfeed, Google's top search results, and the recommendations on whom to follow on Twitter and what to buy on Amazon. They are optimised to prioritise content that will generate more traffic and more profit.

Lotan (2014) warns that "We're not seeing different viewpoints, but rather more of the same. A healthy democracy is contingent on having a healthy media ecosystem. As builders of these online networked spaces, how do we make sure we are optimizing not only for traffic and engagement, but also an informed public? … The underlying algorithmics powering this recommendation engine help reinforce our values and bake more of the same voices into our information streams".

The compounding aspects of this polarisation of opinions in social media have been studied in political science and media and communication studies, e.g. *echo chambers* (Aiello et al. 2012), *filter bubbles* (Pariser 2011), and *voice* (Couldry 2010).

With the Internet spilling over into the built environment, there is an opportune moment. With this book, we collectively call for claiming back the citizen's right to the digital city. The chapters in this book offer a variety of analyses of the unique and largely untapped qualities of the digital city that we want the community of scholars interested in urban interaction design and urban informatics to explore further with us, such as their ability to reach a diversity of citizens, and the—thus far—absence of automated personalisation algorithms. They also offer complementary urban interfaces and civic media channels that may enable non-users of conventional social media to participate in different ways. Can city spaces help us burst the filter bubbles, break out of the echo chambers, and give us a voice?

The citizen's right to the digital city recognises the wealth of knowledge, wisdom, and experiences collectively and privately held by each urbanite. Similar to how Bannon (1992) at the time called for a profound shift in attention "from human factors to human actors" in systems development, more and more commentators these days critique the established hegemony of the engineering and technology-centric epistemology embedded in any one proprietary smart city vision. With this book, we want to contribute to the debate that has started to consider alternative approaches focussing on "smart citizens" and their not just vital, but crucial participation in the city making enterprise (Foth et al. 2011; Townsend 2013; de Waal 2014; Hemment and Townsend 2014; Brynskov et al. 2014).

People have advanced from consumers to co-producers, from stationary office workers to mobile urban nomads, and from passive members of the plebs to active instigators of change. Yet, interaction designers often still refer to them only as "users", and architects and urban planners often still refer only as "city residents" or "building occupants". There is more to it. There is a need to focus on the "life *between* the systems," with a reference to Danish architect Jan Gehl's pivotal book "Life between buildings" (1971/1987). With this volume, we continue the Digital Cities series expanding its repertoire looking at urban interfaces, citizen action, and participatory city making. We trust that the contributions to this book will continue the series that has been seminal in tracing and informing key historic developments at the intersection of digital media and the built environment, from the information superhighway to the smart city.

Started in 1999, the Digital Cities series is the longest running academic workshop series that has rigorously followed the intertwined development of cities and digital technologies. Previous years have seen papers presented at Digital Cities to appear as the basis of key anthologies that we list below. The works presented at the Digital Cities workshop series have also been formative to a diverse set of emerging fields, e.g. urban informatics, smart cities, pervasive computing, Internet of things, media architecture, urban interaction design, and—most recently—urban science (Townsend 2015).

Apart from the Digital Cities series' inclusive nature towards interdisciplinary discussions and debate, the fact that the resulting publications have helped articulate and position issues within this heterogeneous domain is a crucial reason for the longevity and continuity of the Digital Cities biennial gathering. The present edited volume of peer-reviewed chapters is a result of contributions by selected authors that presented at Digital Cities 7 (in conjunction with the 5th Communities & Technologies conference 2011 in Brisbane) and Digital Cities 8 (in conjunction with the 6th Communities & Technologies conference 2013 in Munich), as well as invited colleagues.

We look forward to not only continuing the discussion, but also through our collective work to contributing to betterment of the urban condition, that is, the human condition.

Brisbane, Australia	Marcus Foth
Aarhus, Denmark	Martin Brynskov
Oulu, Finland	Timo Ojala

References

Aiello, L. M., Barrat, A., Schifanella, R., Cattuto, C., Markines, B., & Menczer, F. (2012). Friendship prediction and homophily in social media. *ACM Trans. Web*, *6*(2), 1–33.

Bannon, L. (1992). From human factors to human actors: the role of psychology and human-computer interaction studies in system design. In J. Greenbaum, & M. Kyng (Eds.), *Design at work* (pp. 25–44). Hillsdale, NJ: Laurence Erlbaum Associates.

Brynskov, M., Carvajal B. J. C., Fernández, M., Korsgaard, H., Mulder, I. J., Piskorek, K., Rekow, L., & de Waal, M. (2014). *Urban interaction design: Towards city making*. Neuhausen: BookSprints.

Couldry, N. (2010). *Why Voice Matters*. London: Sage.

de Waal, M. (2014). *The City as Interface: How New Media Are Changing the City.* Rotterdam, NL: NAi010 Publisher. http://www.thecityasinterface.com/.

Foth, M., Forlano, L., Satchell, C., & Gibbs, M. (Eds.) (2011). *From Social Butterfly to Engaged Citizen: Urban Informatics, Social Media, Ubiquitous Computing, and Mobile Technology to Support Citizen Engagement.* Cambridge, MA: MIT Press.

Gehl, J. (1971/1987). *Life between buildings: Using public space*. New York: Van Nostrand Reinhold.

Hemment, D., & Townsend, A. (2014). *Smart Citizens*. Manchester, UK: FutureEverything Publications. http://futureeverything.org/ideas/smart-citizens-publication/.

Jenkins, H. (2006). *Fans, Bloggers, and Gamers: Exploring Participatory Culture*. New York: New York University Press.

Lotan, G. (2014). *Israel, Gaza, War & Data: Social networks and the art of personalizing propaganda*. Retrieved Feb 15, 2015 from https://medium.com/i-data/israel-gaza-war-data-a54969aeb23e.

Pariser, E. (2011). *The filter bubble: What the Internet is hiding from you*. New York, NY: Penguin Press.

Townsend, A. (2013). *Smart cities: Big data, civic hackers, and the quest for a new utopia hard-cover*. New York, NY: W. W. Norton & Company. http://www.smartcitiesbook.com/.

Townsend, A. (2015). *Making sense of the science of cities*. New York, NY: Data & Society Research Institute.

Digital Cities Series

Previous Digital Cities workshops have produced high quality peer-reviewed publications containing selected workshop papers and other invited contributions:

Digital Cities 6 (C&T 2009, PennState)
Foth, M., Forlano, L., Satchell, C., & Gibbs, M. (Eds.) (2011). *From Social Butterfly to Engaged Citizen: Urban Informatics, Social Media, Ubiquitous Computing, and Mobile Technology to Support Citizen Engagement*. Cambridge, MA: MIT Press.

Digital Cities 5 (C&T 2007, Michigan)
Foth, M. (Ed.) (2009). *Handbook of Research on Urban Informatics: The Practice and Promise of the Real-Time City*. Hershey, PA: Information Science Reference, IGI Global.

Digital Cities 4 (C&T 2005, Milan)
Aurigi, A., & De Cindio, F. (Eds.) (2008). *Augmented Urban Spaces: Articulating the Physical and Electronic City*. Aldershot, UK: Ashgate.

Digital Cities 3 (C&T 2003, Amsterdam)
Van den Besselaar, P., & Koizumi, S. (Eds.) (2005). *Digital Cities 3: Information Technologies for Social Capital* (Lecture Notes in Computer Science No. 3081). Heidelberg, Germany: Springer.

Digital Cities 2 (Kyoto 2001)

Tanabe, M., van den Besselaar, P., & Ishida, T. (Eds.) (2002). *Digital Cities 2: Computational and Sociological Approaches* (Lecture Notes in Computer Science No. 2362). Heidelberg, Germany: Springer.

Digital Cities 1 (Kyoto 1999)

Ishida, T., & Isbister, K. (Eds.). (2000). *Digital Cities: Technologies, Experiences, and Future Perspectives* (Lecture Notes in Computer Science No. 1765). Heidelberg, Germany: Springer.

External Reviewers

Alan Chamberlain, University of Nottingham, UK

Jonas Fritsch, Aarhus University, Denmark

Mark Gaved, Open University, UK

Jorge Gonçalves, University of Oulu, Finland

Simo Hosio, University of Oulu, Finland

Kirralie Houghton, Queensland University of Technology, Australia

Marko Jurmu, University of Oulu, Finland

Hannu Kukka, University of Oulu, Finland

Anna Luusua, University of Oulu, Finland

Anna Meroni, Politecnico di Milano, Italy

Peta Mitchell, Queensland University of Technology, Australia

Stig Møller Hansen, Aarhus University, Denmark

Peter Lau Torst Nielsen, Aarhus University, Denmark

Lasse Steenbock Vestergaard, Alexandra Institute/Aarhus University, Denmark

Acknowledgments

We are grateful to Springer for publishing this book and would specifically like to thank our publishing editor, Dr. Ramesh Nath Premnath, for his assistance and guidance through the process. The additional support from Springer staff, particularly Ms. Vani Gopi and Ms. Nishanthi Venkatesan, our book's project coordinator and production editor, has been tremendous. Thank you!

We are honoured that Prof. Saskia Sassen, Chair of the Committee on Global Thought at Columbia University, accepted our invite to contribute the book's epilogue.

Furthermore, we are pleased to see the Digital Cities series continue with instalment #9 having been held recently at the 7th International Conference on Communities & Technologies 2015 in Limerick, Ireland, co-chaired by Michiel de Lange (Utrecht University, The Mobile City), Nanna Verhoeff (Utrecht University), Martijn de Waal (University of Amsterdam, Amsterdam University of Applied Sciences, The Mobile City), as well as two of the book's co-editors Marcus Foth and Martin Brynskov. We look forward to the edited book that is being prepared, as well as to the 10th anniversary instalment of the Digital Cities workshop that will be held in 2017 in conjunction with the 8th International Conference on Communities & Technologies, which will be hosted by the University of Troyes in France.

Finally, in addition to the authors of book chapters themselves, the aforementioned colleagues acted as external peer reviewers who provided invaluable comments and feedback. The standard of critiques that has been provided was superb and played an integral part in shaping the book. Thank you for your critique, thoughts, and contributions.

Marcus Foth
Martin Brynskov
Timo Ojala

Contents

Editors and Contributors

About the Editors

Marcus Foth is founder and director of the Urban Informatics Research Lab, research leader of the School of Design, and professor in interactive and visual design, Creative Industries Faculty, at Queensland University of Technology.

Marcus' research focuses on the relationships between people, place, and technology. He leads a cross-disciplinary team that develops practical approaches to complex urban problems. He adopts human–computer interaction and design methodologies to build engagement around emerging issues facing our cities. Marcus' recent work has examined:

- Urban planning—new approaches to community participation and engagement
- Environmental sustainability—new strategies for energy monitoring in domestic settings
- Food futures—new ideas to rethink eating, cooking, and growing food in the city
- Collaboration and co-working spaces—new aspirations for libraries in the twenty-first century

Marcus has received over $4 million in national competitive grants and industry funding. He received a Queensland Young Tall Poppy Science Award 2013 and was inducted by the planning, design, and development site Planetizen to the world's top 25 leading thinkers and innovators in the field of urban planning and technology.

Marcus has authored and coauthored over 140 publications in journals, edited books, and conference proceedings. He is the editor of the Handbook of Research on Urban Informatics (IGI 2009), coauthor of Action Research and New Media (Hampton Press 2009), coeditor of From Social Butterfly to Engaged Citizen (MIT Press 2011), Eat, Cook, Grow: Mixing Human–Computer Interactions with Human–Food Interactions (MIT Press 2014), and Citizen's Right to the Digital City (Springer, 2016). He was the cochair of the Oxford Internet Institute's Summer

Doctoral Programme 2009, chair of OZCHI 2009, the 5th International Conference on Communities and Technologies (C&T) 2011, and cochair of FoodCHI 2014. He will chair Designing Interactive Systems (DIS) 2016.

Marcus has been invited to give presentations at leading research institutions, including MIT, Harvard, Emerson, Rensselaer Polytechnic Institute, Tsinghua University, University of Oxford, University of Manchester, Helsinki Institute for Information Technology, University of Oulu, IT University of Copenhagen, Ludwig-Maximilians-Universität, Cape Peninsula University of Technology, and the Korea Advanced Institute of Science and Technology (KAIST).
Email: m.foth@qut.edu.au

Martin Brynskov is an Associate Professor in Interaction Technologies at the Department of Aesthetics and Communication at Aarhus University in Denmark. He is also a research fellow at Participatory IT Centre (PIT) and Center for Advanced Visualization and Interaction (CAVI), Director of the Digital Design Lab, co-founder of Smart Aarhus, and Former Director of the Civic Communication group at the Center for Digital Urban Living. He was the General Chair of the Media Architecture Biennale 2012 and 2014.

Timo Ojala is a Professor of Computer Science at the University of Oulu, where he leads the transdisciplinary Urban Computing and Cultures research group. He received his M.Sc. (with honors) and Dr.Tech. degrees in 1992 and 1997, respectively, from the University of Oulu, Finland. He has been the lead investigator of the multidisciplinary UrBan Interactions (UBI) research program, which has conducted a longitudinal exploration of ubiquitous computing systems and applications in authentic urban setting at downtown Oulu. This work has contributed among other things the UBI-hotspots and the annual international UBI Summer Schools for researcher training. Recently, his work has focused on hybrid reality where the real-world city is coupled with a collaborative virtual 3D model of the city regarded as a mirror world. He has authored about 170 papers that have been cited about 15000 times. He has served as the founding co-chair of the 1st International Conference on Mobile and Ubiquitous Multimedia (MUM 2002), the general chair of MUM 2007 conference, the chair of Digital Cities 7, the PC chair or PerDis 2013, the general co-chair of PerDis 2016, and in the program committees of many international conferences and workshops. He is a member of the ACM.

Contributors

Moritz Behrens is an architect and maker, an interaction designer and researcher. In his Ph.D. at the Bartlett School of Architecture, University College London, he is currently looking at social interactions and spatial configurations in technology-mediated urban spaces. Through deploying situated feedback technology in the built environment and in particular connecting to media facades, he explores the potential of collaborative spaces that enhance participation in communities. In the context of his research, Moritz has also been invited to create artistic projects, amongst others for the media facade festivals Viva Cidade 2013 in Sao Paulo, Brazil, the Ars Electronica Festival 2014 in Linz, Austria, and during the Staro Light Festival 2014 in Riga, Latvia. Lately, Moritz won the No.8@Arup competition and implemented an interactive 'Sentiment Cocoon' that explores architectural form, translucent materials and responsive lighting to facilitate social interaction.

Moritz holds an M.Sc. Adaptive Architecture and Computation from the Bartlett School of Architecture and an M.Sc. in Architecture and Urban Planning from the Technical University Berlin. After graduating, he worked for award winning architectural practices in Berlin and as a research associate at the Space Group joining the EPSRC funded 'Screens in the Wild' research project. In collaboration with the Mixed Reality Lab at the University of Nottingham, interactive and networked public displays were implemented in communities around London and Nottingham.
Email: moritz.behrens@ucl.ac.uk

Duncan P. Brumby is a senior lecturer (Associate Professor) in Human–Computer Interaction at University College London (UCL). Prior to joining UCL in 2007, he received his doctorate in Psychology from Cardiff University (2002–2005) and was a postdoctoral researcher in the Computer Science Department at Drexel University (2005–2007). Dr. Brumby has published over 60 peer-reviewed articles on how people interact with computing devices (including 10+ Papers/Notes in the proceedings of the ACM CHI conference: the premier publication venue for Human–Computer Interaction research). Support for Dr. Brumby's work has come from grants provided by the UK's Engineering and Physical Sciences Research Council (EPSRC) and the European Institute of Innovation & Technology. Dr. Brumby is an associate editor for the International Journal of human–computer studies, and has been an associate chair for the ACM CHI conference technical program committee since 2012.
Email: brumby@cs.ucl.ac.uk

Jaz Hee-jeong Choi is the deputy director of the Urban Informatics Research Lab and a senior lecturer in the School of Design, Creative Industries Faculty, at Queensland University of Technology, Australia. Her research interests are in playful technology, particularly the ways in which various forms of playful interaction are designed, developed, and integrated in different cultural contexts. In her previous research, she developed a new conceptual approach to urban sustainability that recognises 'play' as the core of transformative interactions in cities as complex techno-social networks. Her current research explores the intersection of food, design, and technologies, and self-care and mutual aid in urban environments with a particular focus on youth social entrepreneurship. She has collaborated with leading international researchers, published in books and journals across various disciplines, and given invited talks at major international conferences including the opening keynote at the 2010 UNESCO Creative Cities Conference.
Email: h.choi@qut.edu.au

Fiorella De Cindio associate professor in the Department of Computer Science at the Università di Milano, taught programming languages, distributed systems foundations, and software engineering for several years. Since 2002, she has taught a class on virtual communities, now called "Internet-based Social Interaction." Since 2011, she has co-taught a class on "Digital Citizenship and Civic Hackerism" with Andrea Trentini. Early research areas were Petri nets as concurrency theory, concurrent programming languages, participatory design, and computer-supported cooperative work. In 1994, she founded the Civic Informatics Laboratory, and for twenty years. her research has focused on social interactive systems, their design, their implementation, and their deployment in real-life settings; civic participation and deliberation at the urban level; the development of software tools for the purpose. RCM, the Milan Community Network, was her first such real-world project. Since 1998, De Cindio chairs an independent Participatory Foundation. Her activity in the community earned De Cindio the City of Milan's Ambrogino d'Oro award in 2001.
Email: fiorella.decindio@unimi.it

M. Hank Haeusler Dipl.-Ing. (Fh)/Ph.D. (SIAL/RMIT) is discipline director of Computational Design (CoDe) and senior lecturer at Australian School of Architecture + Design at the University of New South Wales, Sydney; board member of the Media Architecture Institute, a non-profit organisation designed to complement the work of established universities and research institutions, and Lead CI of Encircle, an ARC Linkage Grant funded research group investigating responsive transport environments. Previous to his role at UNSW, Haeusler was a Chancellor's Postdoctoral Research Fellow at UTS, Sydney. His Ph.D. thesis 'Spatial Dynamic Media Systems' at SIAL/RMIT University, Melbourne (Supervisor Mark Burry), received a research prize in recognition of the achievement of excellence in a Higher Degree by Research Program assessed in 2008. Haeusler is known as a researcher, educator, entrepreneur and designer in media architecture, digital technology, interaction design and ubiquitous computing, and author of seven books listing publications such as 'Media Facades – History, Technology, Content' (avedition, 2009), 'New Media Facades – A global Survey' (avedition, 2012) and over 30 book chapters and conference papers. He has lectured in Europe, Asia, North America and Australia and was appointed Visiting Professor at Central Academy of Fine Arts, Beijing in 2013.
Email: m.haeusler@unsw.edu.au

Sarah Hartmann is a master's student of Information Science and Language Technology at the Heinrich Heine University and holds a bachelor's degree in the same subject. She is working as a research assistant at the Department of Information Science since 2012 and her research interests are in social media, e-government, and smart cities.
Email: s.hartmann@hhu.de

Larissa Hjorth is an artist, digital ethnographer and professor in the Games Programs, and Deputy Dean, Research & Innovation in the School of Media and Communication at RMIT University, Melbourne, Australia. Since 2000, Hjorth has been researching the gendered and sociocultural dimensions of mobile, social, locative and gaming cultures in the Asia-Pacific—these studies are outlined in her books, *Mobile Media in the Asia-Pacific* (London, Routledge 2009), *Games & Gaming* (London: Berg 2010), *Online@AsiaPacific: Mobile, Social and Locative in the Asia-Pacific region* (with Michael Arnold, Routledge 2013), and *Understanding Social Media* (with Sam Hinton, Sage 2013).

Hjorth has co-edited four Routledge anthologies: *Gaming Cultures and Place in the Asia-Pacific region* (with Dean Chan 2009); *Mobile technologies: from Telecommunication to Media* (with Gerard Goggin 2009); *Studying the iPhone: Cultural Technologies, Mobile Communication and the iPhone* (with Jean Burgess and Ingrid Richardson 2012); and Mobile *Media Practices, Presence and Politics: The Challenge of Being Seamlessly* Mobile (with Katie Cumsikey).

As an artist, Hjorth has been awarded various prestigious awards such as Gyeonggi Creation Center Art Residency (2010), The Australia Council New Work Fellowship (2006), Australia Council Tokyo Studio (2000), Akiyoshidai International Art Village Residency (2002) and the Asialink Seoul Visual Art Residency (2005) as well as gaining over 20 government and corporate grants for her work involving new media in the region. Hjorth has had over 10 solo exhibitions at institutions such as EAF and CACSA, participated in over 50 art exhibitions (such as Yokohama Triennale 2001 with Japanese Internet group, Candy Factory) and curated many cross-cultural projects such as the Japanese and Australian magazine and exhibition project, *gloss* (2002). In 2010, she had a solo exhibition, *Still Mobile*, at Gyeonggi Museum of Modern Art in Korea.
Email: larissa.hjorth@rmit.edu.au

Aylin Ilhan studies Information Science and Language Technology in Düsseldorf, Germany. She is a staff member of the Department of Information Science, Heinrich Heine University Düsseldorf, Germany. Her research fields are ubiquitous cities and social media, including the information behavior of Facebook users. In 2014, Aylin presented the results about her study on New Songdo City in Berlin at SWiF (Student Workshop for Information Science Research) and placed first for the best presentation. In the future, she wants to research further informational cities with regard to their urbanity and their implementation and use of Internet of Things (IoT).
Email: aylin.ilhan@hhu.de

Peter Lyle is a researcher with the Urban Informatics Research Lab, in the School of Design at QUT. His research interests sit at the intersection of computing and social interaction, currently with a focus on food and agriculture. His Ph.D. study is connected to the ‘Eat, Cook, Grow: Ubiquitous Technology for Sustainable Food Culture Attitudes in the City’ study funded by ARC Linkage. In this Ph.D. study, Peter focused on supporting human–computer interaction (HCI) practitioners to engage with food growing communities.

Peter has worked in the Tasmanian IT sector, in a variety of roles from client support and as a network manager, before transitioning to software developer, working on enterprise projects for public and private organisations. From this, he gained a

solid experience base in client interaction. Peter also has a long history of close relations with Apple University Consortium in Australia. As part of his involvement, he has been awarded scholarships to attend a number of Apple University Consortium training sessions focusing on software development for different Apple technologies.
Email: p.lyle@qut.edu.au

Agnes Mainka is a Ph.D. candidate and research assistant at the Department of Information Science at the Heinrich Heine University where she teaches programming and Web design. Her research interests focus on the development of informational cites, e-government, public libraries, and social media.
Email: agnes.mainka@hhu.de

Ian McArthur A Senior Lecturer at UNSW Art & Design, Dr. Ian McArthur, is a hybrid practitioner working in the domains of experimental and speculative multidisciplinary practice, transcultural design collaboration, metadesign and education change.

His work with Brad Miller developing sonifications for responsive interactive media environments, architectural-scale data visualisations and public art installations has been exhibited at VIVID Festival, Sydney (2012, 2014), The Millennium Monument Museum of Digital Art (CMoDA) during Beijing Design Week (2012) and ISEA (Sydney, 2013). His work with Annie Morrad has been presented at the GNARL Performance Research Festival (UK, 2014).

Ian was Program Director of Graphic Design (2001–2003) at La Salle DHU (Donghua University, Shanghai) where he initiated The Collabor8 Project (C8) to foster creative collaboration between China and Australia.

His most recent work is in collaboration with the Chongqing-based innovation platform CQubed, Chongqing University and Priestman Architects (UK/China). CQ.MAD.LAB (2015) is an interdisciplinary studio focused on developing design-led multiscalar precinct development opportunities for the dynamic and rapidly expanding high-rise area of Guanyinqiao in Chongqing.

Supported by the Halloran Trust (University of Sydney), Ian is currently involved in developing participatory design methods using urban screens in public space as diagnostic tools for urban planning.
Email: ian.mcarthur@unsw.edu.au

Tim McGinley Tim's interest in using digital tools to support the participation of non-experts in design began whilst studying architecture in the UK. Later, whilst working as a researcher at the Hyperbody Research Group, TU Delft, he developed a collaborative system architecture for the Protospace Lab to support real-time collaboration of diverse stakeholders in an urban design context. Whilst at Foster + Partners in London, Tim developed parametric design systems for various projects including the design of a new city, the tallest building in the world in construction at the time and Apple's HQ in Cupertino. He received his EngD from the University of Reading, UK, which focussed on the development of the community architecture

framework presented in this chapter, with his supervisor Dr. Keiichi Nakata. As a lecturer at the University of South Australia (UniSA), Tim is currently investigating supporting the design of complex systems in the built environment using architecture frameworks and templates of morphogenetic system development in biology which he calls 'morphogenetic prototyping'. Tim coordinates a Masters of Architecture Studio at UniSA, which he uses to test and extend his research. He lives in Adelaide, Australia, with his partner Aggeliki and their son Kostas, who likes dinosaurs.
Email: Tim.mcginley@unisa.edu.a

Lorena Melgaço is a research assistant on the Digital Neighbourhoods Research project in the School of Architecture, Design and Environment, Plymouth University, UK. Her work is concerned with the articulation of digital technologies, society and urban space. She holds an M.Sc. in International Cooperation and Urban Development from the Technical University of Darmstadt, Germany, and University Pierre Mendes, France, as well as an M.Phil. in Architecture and Urbanism from the Federal University of Minas Gerais (UFMG), Brazil. She is currently a Ph.D. candidate on the topic of 'Spatial practices in socio-technological peripheries: the case of rurban communities in Brazil' at UFMG.
Email: lorena.melgacosilvamarques@plymouth.ac.uk

Christine Meschede holds a bachelor's degree in Mathematics and is currently studying Information Science and Language Technology at Heinrich Heine University. Since 2013, she is working as a research assistant at the Department of Information Science, with interest in smart cities, gamification, and informetrics.
Email: christine.meschede@hhu.de

Hylton Mitchell completed a master's degree in Architecture and Planning in the Department of Engineering and the Built Environment at the University of Cape Town (UCT), where he focused his dissertation on inclusive Internet action of a civil society organisation assisting periphery communities. He holds an Honours Degree in Media Studies from UCT and completed his undergraduate studies at the University of the Western Cape (UWC) in 1996. His research interests lie in the area of developing tools that merge urban planning with digital networks to improve the decision-making associated with developing smart cities in the Global South. He manages an online think-tank @WiFiCities, dedicated to bringing accessible WiFi to African cities, particularly disadvantaged communities.

Hylton has also spearheaded the implementation of a number of mobile and Internet-based tools to promote community safety and development in his local neighbourhood. He serves on Civic Association and Community Safety bodies in his suburb in Cape Town.
Email: hylton_mitchell@yahoo.com

Rena Möhlmann academic career started in 2011 when she began studying Information Science and Language Technology at the Heinrich Heine University in Düsseldorf. She has worked for the Department of Information Science with research fields being ubiquitous cities and social media. For the future, she plans to pursue

a master's degree in computer science and most likely will specialize in databases. When she is not studying or working, she likes to travel the world. So far long trips to New Zealand and Southeast Asia mark the favorite of her travels.
Email: rena.moehlmann@hhu.de

Ingrid Mulder is an associate professor of Design Techniques at ID-StudioLab, Faculty of Industrial Design Engineering, Delft University of Technology. Her expertise is in transformative and social design. As a reader in participatory innovation at the Rotterdam University of Applied Sciences, she has initiated the first Fablab in Rotterdam and the Rotterdam Open Data movement. She also founded Creating 010, a trans-disciplinary design-inclusive research centre enabling citizens, students and creative industry making the future of Rotterdam. Since 2007, she chairs the research programme Meaningful Design in the Connected City, which has an extended research group in both Delft and Rotterdam. Her background is in policy and organization studies (MA, University of Tilburg) and behavioural sciences (Ph.D., University of Twente). In 2005, she headed the evaluation of the first Dutch living lab, "Kenniswijk", currently known as Brainport Eindhoven. Ever since, she has been involved in the interplay between top-down policy and planning with bottom-up participatory innovation. As an expert for the European Commission on Internet of Things and Smart Cities, she advocates participatory bottom-up innovation as a means to empower people in driving social change.
Email: mulderi@acm.org

Keiichi Nakata is associate professor (reader) in social informatics and deputy head of Business Informatics, Systems and Accounting (BISA) at Henley Business School, University of Reading, UK. His main research interests lie in socio-technical systems, in the areas of computer-supported collaborative work, cognitive systems engineering, and information systems. Recently, he has been engaged in research into technology acceptance, enterprise social media and participatory systems, and uses organisational semiotics as one of the theoretical lenses. His recent research projects include semantic information aggregation and sentiment analysis of parliamentary records and development of video-based distributed cognition analysis tools. He obtained his Ph.D. in artificial intelligence from the University of Edinburgh, UK, and M.Eng. and B.Eng. in nuclear engineering from the University of Tokyo.
Email: k.nakata@henley.ac.uk

Nancy Odendaal is a senior lecturer in the School of Architecture, Planning and Geomatics on the Masters Degree in City and Regional Planning, University of Cape Town, South Africa. From 2009 to 2012, she was based at the African Centre for Cities, at the University of Cape Town, where she worked on curricula reform and on building a network of planning schools across Anglophone Africa. She is now on the Steering Committee of the Association of African Planning Schools and incoming chair of the Global Planning Education Associations Network (GPEAN).
URL: http://www.greenmyfavela.org; http://www.learekow.com

Nancy has published on the role of technology in urban change, the use of technological tools in the urban planning process and the relationship between spatial change, service delivery and networked infrastructure. Recent work on planning education and research methodology in Africa has reinforced an interest in the efficacy of planning practice and the role of experiential learning and practice-based research in enabling contextual relevance in planning training. Her current work focuses on infrastructure and spatial change in cities of the Global South.
Email: nancy.odendaal@uct.ac.za

Lea Rekow is director of Green My Favela, an urban restoration project based in the favelas of Rio de Janeiro. Her current research focuses on transdisciplinary practices for reclaiming degraded space in areas where people are living under extreme socioeconomic and environmental stress. Lea is an advising consultant to the Integrated Media program at Cal Arts and the international development action network GlobalCAD. She is also a member of the Institute for Australian Geographers and New York Women in Film and Television. Co-based in New York, Lea has previously held positions as founding director of Gigantic ArtSpace, media director at the former Center for Peace and Human Security, adjunct professor of media and communications at Pratt Institute, and cultural advisor to Advance. She was also executive director at the Center for Contemporary Art in Santa Fe and research fellow at the Center for Art and the Environment. Lea has also acted as an advisor to the European Urban IxD program, NYFA, SVA, Parsons, Amnesty International, the MacArthur award, and the LMCC. Her work involves exchanges and collaborations that connect with a diverse range of people, human security concerns, and environmental issues, including the legacy of energy and mining on Navajo Nation, and the civil war and its impact on the ethnic minorities in Burma.
Email: learekow@gmail.com

Ava Fatah gen. Schieck is an architect, educator and researcher. She is an associate professor (reader) at the Bartlett School of Architecture, University College London, where she runs two modules: the ‘Body as Interface’ and the ‘City as Interface’ on the M.Sc. Adaptive Architecture and Computation programme. Her research work has developed over the last 14 years through her teaching and research positions she held at UCL and has held a number of the UK Engineering and Physical Sciences Research Council (EPSRC) funded awards. Ava is the principal investigator of the ‘Screens in the Wild’ project, which aims to explore the potential of networked urban screens for communities and culture, and offers a unique longitudinal living laboratory environment that explores urban digital interactions (situated, mobile and networked) since 2011.

She has lectured extensively and has more than 70 publications in books, scientific journals and conference proceedings, mainly in the area of Media Architecture, Urban Interaction Design, and Ubiquitous Computing. Ava serves on programme committees of international conferences such as ‘Rethinking the Human in IT-Driven Architecture’, ‘Pervasive Displays’, ‘Digital Cities’ (on Communities and Technologies) and ‘Media City’. She is a chair on the academic conference and a member of the organizing committee on Media Architecture Biennale (2012, 2014 and 2016).
Email: ava.fatah@ucl.ac.uk

Saskia Sassen is the Robert S. Lynd Professor of Sociology and Chair, The Committee on Global Thought, Columbia University (www.saskiasassen.com). Her new book is *Expulsions: Brutality and Complexity in the Global Economy* (Harvard University Press 2014). Recent books are *Territory, Authority, Rights: From Medieval to Global Assemblages* (Princeton University Press 2008), *A Sociology of Globalization* (W.W. Norton 2007), and the 4th fully updated edition of *Cities in a World Economy* (Sage 2012). Among older books is *The Global City* (Princeton University Press 1991/2001). Her books are translated into over 20 languages. She is the recipient of diverse awards and mentions, including multiple doctor honoris causa, named lectures, and being selected as one of the top global thinkers on diverse lists. Most recently, she was awarded the Principe de Asturias 2013 Prize in the Social Sciences and made a member of the Royal Academy of the Sciences of Netherland.
Email: sjs2@columbia.edu

Jan Seeburger completed his Ph.D. at the Urban Informatics Research Lab in 2012. His research investigated how ICT can be utilised to create, influence, and evaluate user experiences in urban public places mediated through mobile devices. Several mobile phone and urban screen applications have been designed, developed, and deployed in a real-world context within his Ph.D. study.

Before commencing his Ph.D., he wrote his Master Thesis at the Research Centre for Computer Science (FZI), Karlsruhe, Germany, worked on projects for SAP Research in Darmstadt, Germany, and SAP Research Brisbane, Australia, and wrote his Honours' Thesis at the Research Centre for Intelligent Media, Furtwangen, Germany.

He is currently employed as a research fellow at the Urban Informatics Research Lab investigating opportunities of mobile applications, social media, and urban informatics to increase the loyalty rate of young male blood donors.

Jan can be reached on Twitter at @Jan_See. His private website is http://studio2k.biz/
Email: jan.seeburger@gmail.com

Wolfgang G. Stock is the head of the Information Science Department of the Heinrich Heine University Düsseldorf. His research projects include information retrieval, knowledge representation, informetrics, and studies on information markets as well as on informational cities. Recently, he published, in cooperation with Mechtild Stock, the "Handbook of Information Science" (De Gruyter Saur, Paperback Ed. 2015).
Email: stock@phil.hhu.de

Stefano Stortone post-doc researcher at the Università di Milano Civic Informatics Laboratory, holds a Ph.D. in politics from the Università Cattolica del Sacro Cuore in Milan, where he also did his undergraduate work, and an M.A. in democratic studies from the University of Leeds. His research focuses on democratic theory, civil society, and political participation that aims to overcome the crisis of political representation and representative democracy. He is currently studying the use of ICT for public deliberation and democratic decision-making processes. He heads

the Center for the Study of Participatory Democracy, a nonprofit association that promotes participatory practices and develops participatory budgeting projects for local government throughout Italy, as a step toward experimenting with new, ICT-driven forms of democracy and democratic institutions.
Email: stefano.stortone@unimi.it

Dian Tjondronegoro is an associate professor in mobile multimedia: currently leading the Mobile Innovation Lab and contributing towards the research and development for three Cooperative Research Centres (CRC): Smart Services CRC (smart video optimisation for mobile and multichannel delivery), Rail Innovation CRC (mobile technologies for delivering real-time transit information and mobile payment), and Young and Well CRC (mobile tools for promoting health in young people, including awareness, treatment, and prevention). Dian has published over 88 academic publications in mobile multimedia field and is the current coordinator of Continuing Professional Education (CPE) courses in the mobile computing series, as well as the undergraduate and postgraduate courses in introduction to mobile computing and mobile apps design and development.
Email: dian@qut.edu.au

Martin Tomitsch is Associate Professor in Design at the University of Sydney in Australia, where he teaches interaction design and human–computer interaction. He is the director of the Design Computing Program and Head of the Design Lab, an interdisciplinary research group within the Faculty of Architecture, Design and Planning. He holds a visiting lecturer appointment at the Vienna University of Technology Research Group for Industrial Software (INSO) in Vienna, Austria, is research supervisor for the Central Academy of Fine Arts (CAFA) Media Architecture R&D Lab in Suzhou, China, founding member of the Media Architecture Institute (MAI) and NSW state cochair for the Computer-Human Interaction Special Interest Group (CHISIG) of the Human Factors & Ergonomics Society of Australia (HFESA). Before entering academia, he worked as user experience designer in large software and IT projects.
Email: martin.tomitsch@sydney.edu.a

Katharine S. Willis research over the last decade has explored the effects and implications of digital networks on the experience and design of urban space and place. She has authored and edited over forty publications on these themes. Her most recent book is Netspaces: Space and Place in a Networked World (Ashgate Press 2015). Other recent books include: 'Locative Media: Multidisciplinary Perspectives on Media and Locality' (co-edited, Transcript Press 2013), 'Shared Encounters' (co-edited, Springer 2010) and 'Mediacity: situations, practices and encounters' (co-edited, Frank and Timme 2009). Katharine Willis trained as an architect with a Masters in Architecture from the Bartlett, University College London. She holds a Ph.D. from the Bauhaus University of Weimar, Germany, where she was an EU Marie Curie Fellow in the MEDIACITY research project. Since 2011, she has been based in the School of Architecture, Design and Environment at University of Plymouth, UK, where she is associate professor (reader) in Digital Environments.
Email: katharine.willis@plymouth.ac.uk

Part I
Urban Interfaces

Digital Design Interventions for Creating New Presentations of Self in Public Urban Places

Jan Seeburger, Marcus Foth and Dian Tjondronegoro

Abstract During everyday urban life, people spend time in public urban places waiting for specific events to occur. During these times, people sometimes tend to engage with their information and communication technology (ICT) devices in a way that shuts off interactions with collocated people. These devices could also be used to better connect with the urban space and collocated people within. This chapter presents and discusses the impact of three design interventions on the urban user experience enabling collocated people to share lightweight, non-privacy-sensitive data in the urban space. We investigate and discuss the impact on the urban experience under the notions of people, place, and technology with an emphasis on how the sharing of non-privacy-sensitive data can positively transform anonymous public urban places in various ways through anonymous digital augmentations.

Keywords Urban informatics · Mobile phones · Public displays · User experience · Urban experience · Public space

J. Seeburger · M. Foth (✉) · D. Tjondronegoro
Urban Informatics Research Lab, Queensland University of Technology, Brisbane, Australia
e-mail: m.foth@qut.edu.au

J. Seeburger
e-mail: jan.seeburger@gmail.com

D. Tjondronegoro
e-mail: dian@qut.edu.au

M. Foth et al. (eds.), *Citizen's Right to the Digital City*,
DOI 10.1007/978-981-287-919-6_1

1 Introduction

As the majority of the world's population now lives in urban environments (United Nations 2008), much attention has recently been paid to the research field of urban computing and urban informatics which promise to solve some of the resulting problems that modern cities face through the densification of the urban space such as providing more effective transportation or improving the sustainability of cities. Urban computing is defined as *"the integration of computing, sensing, and actuation technologies into everyday urban settings and lifestyles"* (Kindberg et al. 2007). While urban computing research is mostly focussing on the technology itself, the emerging research field of urban informatics is investigating the impact of technology on people in urban environments (Foth 2011). Urban informatics is defined as *"the study, design, and practice of urban experiences across different urban contexts that are created by new opportunities of real-time, ubiquitous technology and the augmentation that mediates the physical and digital layers of people networks and urban infrastructures"* (Foth et al. 2011). One opportunity to utilise real-time ubiquitous technology in urban environments could be to augment the public urban space with a digital layer containing information about, or left behind by, collocated urban dwellers (Seeburger 2012a; Seeburger et al. 2012b).

The major uptake of social media in combination with the widespread use of mobile devices equipped with a constant connection to the World Wide Web and GPS functionality gave rise to new kinds of mobile applications enabling people to connect to their social circle while on the move. During everyday urban life, many people often spend time in public urban places waiting for specific events to occur. While waiting, those people sometimes tend to use their ICT devices to create a cocoon while in public shutting off interactions with collocated people (Bassoli et al. 2007; Crawford 2008; Mainwaring et al. 2005). On the other hand, location-based services, which deliver content according to the location of the user (Junglas and Watson 2008), enable urban dwellers to retrieve information about nearby places. Gordon and de Souza e Silva (2011) introduce the term *NetLocalities*, stating that *"location-aware mobile technologies can change the way we experience both, physical and digital spaces by configuring a new hybrid space, which is composed by a mix of digital information and physical localities"*.

This chapter reports on our research efforts studying the impact of utilising location-aware mobile technologies on the experience of physical spaces. Throughout a three-year study, several digital artefacts have been deployed and evaluated in the urban space enabling anonymous collocated people to collaboratively augment the urban space (Seeburger 2012b). Each digital artefact has been evaluated against the user experience it helped create. The term "user experience" has had many instantiations and variations in usage in the literature (Bargas-Avila and Hornbæk 2011; Law et al. 2009; Hassenzahl and Tractinsky 2006). We have previously critically reviewed those in order to establish a framework for designing and evaluating user experiences (Seeburger et al. 2012a) that has been applied for the purpose of this study. This chapter presents an analysis that goes beyond the usability and artefact-centric notion of user experience, that is, interacting with the digital artefacts, and instead focusses on discussing the wider impact on the urban experience.

In this chapter, we try to answer the question how the establishment of mobile-mediated interactions and digital augmentations can influence the experience of people in public urban places. Our discussion draws on research informed by urban sociology, psychology, and urban design to provide insights into how mobile human–computer interaction and urban informatics (Foth et al. 2011) design interventions can create urban user experiences that go beyond interacting with a product and digitally augment and amplify the collective sociality in public urban spaces. We do not advocate to immediately subscribe to the goal of forming ties directly with collocated strangers by "making people talk to each other". Instead, we consider the affordances of the digital for an artful integration that dovetails with place, performativity, and interfaces to enhance sociality outside the groups that people already socialise with via social media.

The remainder of this chapter is structured as follows: the next section presents and discusses the three design interventions and the interactions and augmentations they create. Afterwards, a detailed discussion on how such applications can have an impact on the user experiences in public urban places with regard to people, place, and technology is presented, followed by concluding remarks and future research directions.

2 Urban Digital Design Interventions

The goal of this chapter is to provide a critical and theoretically grounded synthesis of our deployment of three mobile and screen-based applications, which have been designed to enable collocated people to anonymously exchange lightweight data in the urban space (Seeburger 2012a; Seeburger et al. 2012b). Figure 1 visualises the user interfaces of the designed applications. The iOS application Capital Music is depicted on the left, a PlaceTag and the respective interactive mobile website in the middle, and YourScreen on the right side of the image, whereas the

Fig. 1 Three design interventions aiming to positively influence the experience in urban public places

top image shows the screen waiting for submissions and the bottom image shows the user interface while playing a video submission. The following subsections describe each design intervention in more detail.

2.1 Capital Music

Capital Music is a mobile application designed for iOS devices, enabling application users to listen to their music as usual but also visualising album artworks of songs currently played in vicinity. Users are able to anonymously send and receive "like" and text messages based on currently played song only utilising nicknames. The user interface visualised on the left side of Fig. 1 shows a mosaic of album artworks, whereas each artwork represents a collocated user. Tapping one of the artworks reveals more information about the song choice such as album, artist, title, and genre.

A paper-based study utilising Post-it notes evaluated the concept of anonymous music sharing and guided the design and the development of Capital Music (Seeburger et al. 2010). A user experience study utilising the Wizard of Oz method to simulate a real-world environment has been conducted and described in detail (Seeburger et al. 2012b). In summary, Capital Music raised awareness towards collocated people and piqued users' curiosity in terms of who is listening to what in the physical space. Capital Music has been described as a voyeuristic tool to peak into people's music listening habits and aids in reconsidering perceptions towards collocated people. As a design research intervention, it continues an inquiry into the impact of personal digital devices and ubiquitous computing in the urban environment, particularly the individual consumption of music (cf., Bull 2012; Gopinath and Stanyek 2014). Sharing, recommending, and discovering music in combination with anonymous social user interaction has been perceived as a fun and entertaining activity while using public transport.

The peer-to-peer approach employed by Capital Music enables users to exchange lightweight data in the form of metadata about currently played songs with other users who are in the same place at the same time. A radius slider in the user interface allows users to control the reach of the song updates they receive. A user can be in multiple location-based peer-to-peer networks and thus exchange song information as visualised in Fig. 2.

Capital Music interactions and augmentations are only visible to application users in a "right here and right now" approach, whereas the album artworks being exchanged are not persistent and therefore not visible for later users.

2.2 PlaceTagz

PlaceTagz are QR codes printed on stickers that are linked to an online message board enabling collocated users to interact with each other over time resulting in a

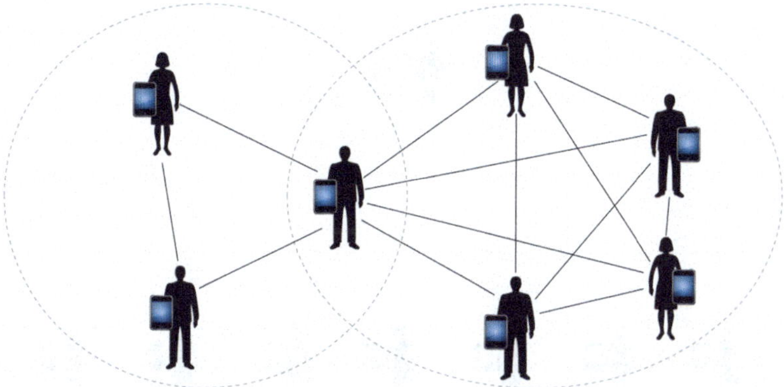

Fig. 2 Capital music interactions visualised

place-based digital memory. As visualised in the middle of Fig. 1, a PlaceTag sticker only contains the QR code and an icon without explicitly revealing purpose or destination. After scanning a PlaceTag with a mobile phone, users are redirected to a website containing location-based information and questions—enabling users to leave a comment for later users. The stickers have been mostly deployed in places where people spend time on their own, the place is mostly used for only a short period of time, or people are resting and are waiting for a specific event to occur. Examples are public toilets, park benches, public workstations in a library space, or coffee shop counters.

A content analysis of 121 received comments and semi-structured interviews with people who submitted comments has been conducted (Seeburger 2012a). Summarised, decontextualised QR codes with means for social user interaction and location-based digital representations were perceived as novel, intriguing, and interesting, and stimulated user responses relating to collocated people, place, and the PlaceTagz technology itself.

Figure 3 visualises a conceptualised view of how one PlaceTag enables interactions and augmentations in the urban space. PlaceTagz enable interactions for collocated people over time resulting in persistent augmentations and interactions visible for future users who scan the QR code.

2.3 YourScreen

YourScreen is a public urban screen application deployed at a bus stop allowing users to collectively watch online videos while waiting for the bus. As part of this design intervention, the bus stop screen has been equipped with a user account from the micro-blogging system Twitter and a fan page on the social networking service Facebook. Urban dwellers can use these services to submit videos selected from the online video service YouTube utilising their mobile devices.

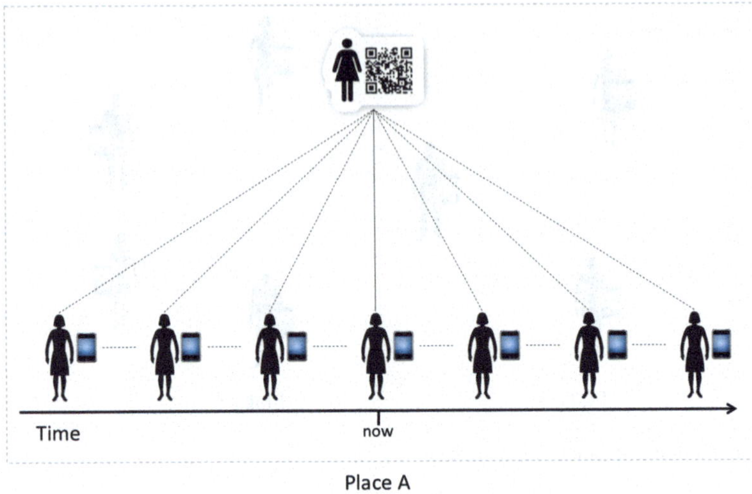

Fig. 3 PlaceTagz interactions visualised

The received video submissions and the conducted user study showed that urban dwellers submitted videos out of curiosity, to overcome boredom, and observe reactions of collocated people. YourScreen has been perceived as awesome, fun, pretty cool, amazing, brilliant, and ingenious application and urban dwellers submitted videos they thought other collocated people might like. Additionally, urban dwellers not present at the screen location submitted videos for people located at the screen to share their favourite videos.

Figure 4 shows a conceptualised view of the interactions and augmentations enabled through YourScreen. YourScreen enables urban dwellers to share digital augmentations in the form of videos with people who are at the same place at the same time. While the interaction with the public urban screen privately takes place on the mobile phone, the resulting interaction is visible for collocated people who spend time at the bus stop. Like Capital Music, the augmentations are not persistent and only visible at the time of interaction.

3 Impacting the Urban Experience

The following subsections discuss the impact of anonymous mobile-mediated interactions on the urban experience utilising urban informatics (Foth 2011) guiding triad of people, place, and technology as analytical pillars. However, the division between people, place, and technology is often overlapping and cannot always be clearly distinguished. People collocated in public urban places and the social interactions which occur are experienced as the "filling" of space (Lechner 1991). Therefore, we discuss the impact on the urban experience as perceived by the individual urban

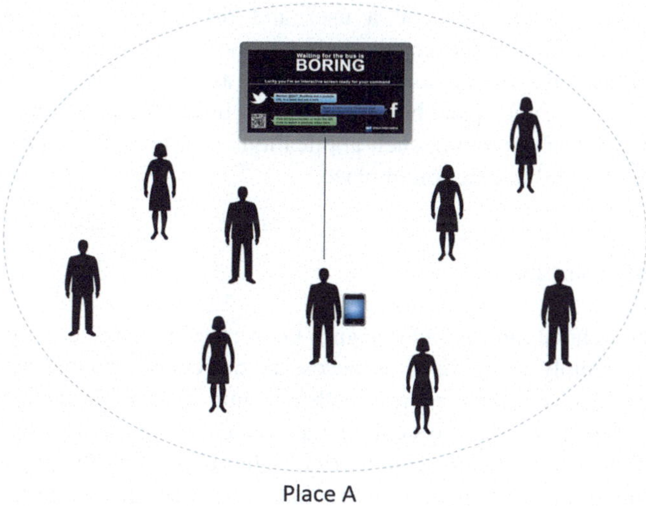

Fig. 4 YourScreen interactions visualised

dwellers in the *people* subsection. The *place* section analyses the urban experience of collocated people in place informed by Goffman's theory of *The Presentation of Self in Everyday Life* (Goffman 1959) as well as the activities that occur in a place. While the *place* section discusses collocated people in space, the *people* section elaborates the concepts of boredom and curiosity as well as the resulting awareness of an urban dweller in relation to the design interventions discussed in this chapter.

Public urban spaces are experienced through technologies such as streets, cars, or buildings, and location-aware technologies contribute to an additional digital layer of a technologically mediated urban experience (Gordon and de Souza e Silva 2011). The *technology* section discusses how design interventions enabling mobile-mediated interaction should consider and align towards the concept of anonymity in public urban places.

Each design intervention has been studied with regard to the user experience in terms of how people feel about interacting with such services. A summary of the user experience that our study revealed has been provided in the previous section. The following discussion goes beyond the user experience of the design interventions and instead investigates from a metaperspective the impact of anonymous mobile-mediated interactions on the experience of everyday life as experienced in public urban places.

3.1 Place

The three design interventions are explicitly location based and focus on the collocated people within. As shown in Fig. 2, Capital Music has been explicitly designed for interactions between collocated people, whereas YourScreen interactions (Fig. 4) are between a user and the artefact. The result of the YourScreen interaction, however, is visible to other collocated people. On the other hand, PlaceTagz interactions (Fig. 3) are asynchronous, i.e. designed for people who use the same space but at a different time. The following subsections discuss the impact of deploying such applications in urban public places in terms of activities and social interaction in place.

3.1.1 Activities and Place

According to architect and urban designer Gehl (1987), activities in public urban places can be simply categorised as necessary activities, optional activities, and social activities. *Necessary activities* such as commuting to work or buying groceries do not depend on the physical characteristics of the urban environment and take place all year long. *Optional activities* highly depend on the physical characteristics of the urban environment and include activities such as going for a walk or relaxing in a park. *Social activities* depend on collocated people and include activities such as greeting other people, conversations, or even just passive contacts. Gehl (1987) states that optional activities will occur to a greater extent if the physical characteristics of a public urban space are inviting. Social activities are *resultant* activities because they evolve from necessary and optional activities. The amount of necessary and optional activities influences the amount of social activities. If a public place is perceived as inviting, people will allow more time for necessary activities and more people will use a public urban place for a variety of optional activities. Gehl states that social activities often occur spontaneously out of the variety of activities that occur within a public urban place.

Gehl originally published the first edition of his book (1987) in 1971 and focused on the urban planning and architecture aspects of public urban places. Nowadays, the widespread use of location-aware and Internet enabled device can impact on the way that urban dwellers experience the physical environment (Gordon and de Souza e Silva 2011). The possibility to digitally interact with collocated people in an anonymous way can change the relationship between the above-described chain of necessary, optional, and social activities in urban public places. This topic considering place, performativity, and interface has also been discussed and analysed by Bull (2012), Farman (2011), and others.

The designed digital artefacts presented in this chapter provide means for digitally mediated social activities, while a necessary activity occurs. While Gehl states that social activities can occur from necessary activities, he argues that optional activities are increasing the likelihood of social activities. Having ICT devices used in public urban places, however, can impact on the relationship between necessary, optional, and social activities. For example, collocated urban dwellers can digitally interact with each other in the form of exchanging metadata of currently played songs using the Capital Music application while taking part in the necessary activity of commuting. Exchanging such presence data can be considered as a digital alternative to what Gehl (1987) terms the social activity of passive contact. Thereby, using ICT devices to interact with collocated people creates a virtual optional activity that may not depend on the physical characteristics of the space, yet it can result in social activities. Additionally, physical artefacts, which enable digital interactions and augmentations such as urban screens or QR codes, are changing the physical characteristics of a public urban place. While such changes to the physical environment do not necessarily make public urban places more inviting, they do provide—as discussed above—new opportunities for optional activities that can facilitate social activities. Overall, the digital layer created by the design interventions deployed in our study can create new optional activities resulting in digitally mediated social activities. Notwithstanding physical characteristics such as the climate at a public urban place that are some of the influencing factors of how inviting a public urban place can be (Gehl 2010), the digital layer also has the ability to make a particular place more inviting.

3.1.2 Social Interaction in Place

Our study brings Goffman's theory on *The Presentation of Self in Everyday Life* (1959) to bear as a theoretical framework for discussing the design interventions with regard to social interactions in public urban places. Goffman uses the metaphor of theatrical performances to describe the process of social interaction in public urban places. He describes how the actions of actors are based on social performances to *give* (explicit information such as the words spoken) and *give off* (implicit information such as body language and fashion) the right impressions to the audience, which in turn is also performing for the actor. It has to be noted that digitally mediated interactions lack the richness of face-to-face communication and has different kinds of implicit, *given off* information one consciously or unconsciously shares (Miller 1995). Rather than *giving off* information such as body language or tone of voice, digitally mediated interactions usually only contain little implicit information such as style of writing (Miller 1995) or—in the case of the design interventions discussed in this chapter—the intention behind the content that a user shares.

Figure 2 shows a conceptualised view of the mobile-mediated interactions created by Capital Music. As these interactions are similar to a face-to-face setting, we found Goffman's theory partly applicable to analyse the behaviour of application users utilising his key concepts. The Capital Music user interface provides the stage where the performances take place. *Props* used in the *setting* are the songs a user selects to share with others and constructs the *front*, which defines the situation for other users, the observers. In Goffman's terms, a Capital Music user might listen to a specific song to *give* a recommendation to others but also *give off* the impression that his or her music taste is, for example, non-mainstream, demonstrating a certain kind of identity and belonging what furthermore contributes to the manifestation of the *front*.

Introducing design interventions as discussed in this chapter results in a digital presentation of self added on top of the physical presentation of self in everyday life. Research has been conducted on the presentation of self with electronic communication such as emails and personal websites (Miller 1995; Schau and Gilly 2003), mobile phone calls (Fortunati 2005), Facebook (Birnbaum 2008), and the presentation of place through location-based services (Sutko and de Souza e Silva 2011). The design intervention discussed in this chapter is presenting a unique case of the digital presentation of self, because they are designed to anonymously exchange non-privacy-sensitive data between collocated people. Interacting with Capital Music while on a bus results in additionally maintaining and managing the digital presentation of self on top of the physical presentation of self. However, due to the anonymity of the design interventions, the digital presentation is decoupled from the physical representation resulting in the need to manage two fronts simultaneously. The physical front while using Capital Music would be commuting with public transport and waiting for the bus while interacting with YourScreen. Our observations (Seeburger et al. 2010) show that these activities mostly involve managing the information that is implicitly *given off* rather than explicitly *give* information in the form of social user interaction, since people do not directly interact with each other while waiting for a specific event to occur (waiting for the bus at the bus stop, sitting in the bus, and waiting till the bus arrives). People are now given tools to create and perform in their desired front using the data they share to *give off* information. Additionally, Capital Music allows users to directly *give* information through the built-in messaging functionality. From this analytical point of view, Capital Music and YourScreen provide attractive means to experience well-known urban scenarios such as waiting for or commuting with public transport by being able to create *fronts* and observe the theatrical performances and reactions from collocated people—with the onus of direct face-to-face interaction being completely optional.

As shown in Fig. 3, PlaceTagz are in contrast to Capital Music and YourScreen mostly used in isolation, enabling an urban dweller to read previous interactions and explicitly *give* information for successive collocated people. PlaceTagz content can also contain implicitly *given off* information (Miller 1995). Due to the nature of PlaceTagz interactions, the digital representation of a QR code can be seen as the digital stage where the theatrical performance occurs. While interacting with PlaceTagz, a user is in the *back* where no other performers are present and can prepare their performance. After submitting the text comment, the performance in the form of a text comment is then added to the digital stage demonstrating the user's *front* comprising *given* and *given off* information.

It has been argued that talking on a mobile phone while spending time in a public urban place results in the revelation of the *backstage* and the exposure of various roles which might differ to the presented *front* (Fortunati 2005). The widespread adoption and use of location-based services such as Foursquare result in a presentation of place, "a multiplicity of agents giving and giving off impressions—impressions that collectively become impression of a place" (Sutko and de Souza e Silva 2011). While PlaceTagz might also contribute to the presentation of place, we showed that the interactions generated through the discussed design interventions enable new means to create and be playful with digital fronts to anonymously interact with collocated people resulting in a digital presentation of self in place.

3.2 People

While conducting user studies with the described digital artefacts, two themes could be identified how such design interventions integrate with and impact on urban life. First, the described design interventions help to relieve boredom while waiting, and second, they pique users' curiosity. Boredom and curiosity can be interrelated, whereas boredom can be the fertile ground for curiosity (Loewenstein 1994). The research presented in this chapter shows that curiosity can cause raised awareness towards collocated people.

3.2.1 Boredom and Waiting

Boredom has been defined as "an unpleasant, transient affective state in which the individual feels a pervasive lack of interest in and difficulty concentrating on the current activity" (Fisher 1993). The term "boredom" has emerged in the eighteenth century following the "development of the notion of leisure and the increase in what has been called leisure time" and the "increasing emphasis on the individual, particularly the greater concentration on the self" (Conrad 1997). The mobile phone, which contains personal contacts and multimedia data, has been described as a personal device that can be seen as an extension of the human body (Townsend 2000). With the uptake of social networking services such as Facebook, an urban dweller's extended social circle is always available in their pocket. Research has also shown that mobile phones are used as a social utility to relieve boredom through socialising (Wei and Lo 2006). On the other hand, it has been argued that there is a positive value in being bored and people should disconnect from technology and actively embrace in boredom to allow the brain to reset itself and enable new ideas to flow (Bell 2011). Humans experience boredom when being disconnected from what is going on or when they are understimulated and there is nothing to do (Conrad 1997). The design interventions described in this study all aim at being used while idle, waiting, and being bored.

The research conducted with the YourScreen application showed that urban dwellers mostly engage with their mobile devices or observe the streets while waiting for the bus. The opportunity to watch an online video while waiting for the bus gives people something novel to do in their idle time. PlaceTagz have been specifically deployed at locations where people have to wait. For example, an interview participant told us that he was bored waiting for an event to start. He then discovered the PlaceTag, scanned the QR code, and left a comment for future visitors in order to keep the digital conversation going. An observation conducted for the Capital Music study revealed that people mostly either read, listen to music, interact with their mobile device, or stare out of the window if they travel alone. Participants for the user experience study explained that travelling with public transport can be a boring activity and that the social interaction mechanisms integrated into the application can help to transform commuting into a fun experience (Foth and Schroeter 2010). Additionally, discovering the music played by collocated people has been described as a fun and entertaining activity while spending time in public transport.

All three design interventions give people the possibility to escape the boredom of mundane activities by providing the possibility to interact with a digital artefact. As explained at the beginning of this subsection, boredom is tackled from both sides in terms of understimulation and disconnection: first, the digital artefact gives people something to do, and second, it provides information to connect to the place and the people within.

3.2.2 Curiosity

According to research in psychology, curiosity is an "intense pleasant experience" that has been defined as "a positive emotional-motivational system associated with the recognition, pursuit, and self regulation of novel and challenging opportunities" (Kashdan et al. 2004). Curiosity can result from the deprivation in cognition through a gap in a person's knowledge (Loewenstein 1994). Russian psychologist Ivan Pavlov already discovered during his experiments on conditioned responses that dogs would naturally investigate any unusual external object or light (Pavlov 1927). Human beings are curious in a way that they try to possess information about collocated people before they initiate social interaction (Goffman 1959).

Curiosity has been categorised by Berlyne along two dimensions: perceptual and epistemic curiosity as well as diverse and specific curiosity (Berlyne 1954). Perceptual curiosity is brought about by novel stimuli and reduced by continued exposure; epistemic curiosity is based on the desire of gaining knowledge. Diverse curiosity is grounded in seeking stimulation while being bored, and specific curiosity describes the desire for a particular piece of information. Figure 5 visualises the dimensions of curiosity as categorised by Berlyne but also mapping the design interventions onto the two-dimensional chart. The data from the conducted studies show that when used for the first time, all three applications are mostly situated within the perceptual/diverse corner of the curiosity chart. As described in the previous section, the digital artefacts give people the possibility to escape the boredom of mundane everyday activities. Before using one of the systems, the curiosity is based on the desire to escape boredomness (diverse curiosity) in combination with the novel digital artefacts (perceptual curiosity) in the urban space. For example, a user is bored while waiting at a coffee shop for his order when she discovers a PlaceTag, whereas the QR code symbolises the availability of digital information but does not physically reveal its destination. In the case of the YourScreen application, a user might be bored while waiting for the bus but discovers the urban screen application and the novel possibility to send and watch a YouTube video.

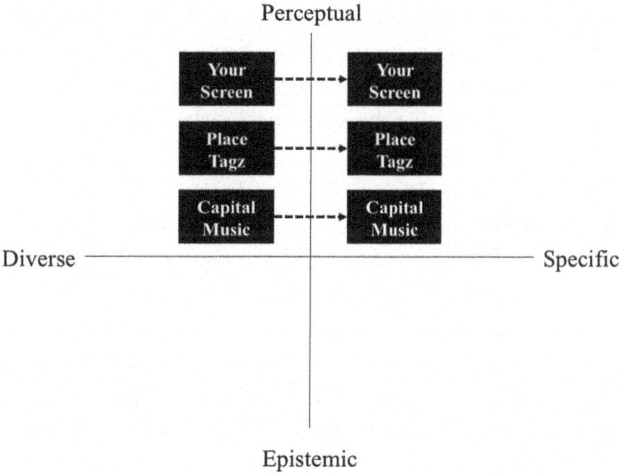

Fig. 5 Design interventions mapped onto the two dimensions of curiosity (Berlyne 1954)

While interacting with one of the systems, the curiosity categorisation shifts from the diverse to the specific curiosity. For example, after scanning the QR code of the PlaceTag, the user is puzzled and surprised about its purpose and background of the digital information. Our data showed that in case of the YourScreen application, users were mostly curious if it actually works and how long it takes until the video is shown on the screen as well as how and to what extend collocated people might react to the video submission. While using Capital Music, people are curious about who in the physical space is listening to what song as visualised in the user interface.

As described by Berlyne (1954), it must be noted that the perceptual curiosity towards the novelty of the application is decreasing with continued exposure. Figure 5 shows the curiosity before and during initial use. Established users, however, may lose perceptual curiosity and switch back and forth between diverse and specific curiosity.

Curiosity has been discussed under the umbrella of interaction design suggesting to create a "zone of curiosity" through hiding small parts rather than large parts of information (Anderson 2011). PlaceTagz hide their content and purpose in the physical space, but scanning them will reveal the information withheld. On the other hand, the Capital Music application never reveals who in the physical space is listening to what song visualised in the user interface. Users are not able to break through the "zone of curiosity", and only digitally mediated social interaction with collocated users will reveal the hidden information.

With the above discussion in mind, it seems that urban informatics design interventions vary on the duration and level of curiosity they might pique. The curiosity of YourScreen is mainly based on its technical functionality. The curiosity of PlaceTagz is based on their contextless appearance and their hidden content. The curiosity of Capital Music is based on the people collocated in the same place. Our studies show that curiosity towards technology is easy to satisfy and decreases upon use, whereas curiosity towards people is a long-lasting experience that in the case of Capital Music can never be fully exhausted. As shown within this study, curiosity can be a major component of the user experience that urban informatics design interventions might create. Being curious is thus related to a positive experience in relation to the self, the world, and the future (Kashdan et al. 2004).

3.2.3 Awareness

The research conducted with Capital Music, YourScreen, and PlaceTagz showed that the technologies created as experiments for our study raise the awareness towards people who are in the same space at the same time or have been in the same place previously. The Oxford dictionary defines awareness as "knowledge or perception of a situation or fact". The experiences created by Capital Music, YourScreen, and PlaceTagz can influence the perception of users towards people collocated in place. Goffman's concept of civil inattention describes how strangers acknowledge and are aware of one another in a public urban space such as an elevator, but do not require or expect to directly interact with each other (Goffman 1966). While people are generally aware of each other when spending time in a public urban place, the exchanged information through interacting with one of the design interventions increases this awareness through combining digital information with physically present people or place. Previous research has shown that location-based social networks such as Foursquare can raise the awareness towards friends nearby (Sutko and de Souza e Silva 2011). However, location-based social networks have also been criticised that they mostly connect people who already know each other (Crawford 2008). Our study tried to broaden the scope of appeal to also include the civil inattention sphere between anonymous people. We found that providing design interventions tailored for idle situations in public urban spaces, and also piquing a user's curiosity, can result in new forms of expression of self as well as raise awareness towards collocated people in place through these new forms of digital self-expression.

3.3 Technology

The three design interventions have been designed to be used without utilising user profiles or subscriptions, and enable urban dwellers to anonymously share lightweight data. This subsection discusses the concepts of anonymity and lightweight content sharing in urban public places.

3.3.1 Anonymity

Anonymity is a social construct brought about by the presence of other people and the desire to not being identifiable to them (Marx 1999). Urban life is mostly anonymous, and the vast amount of people inhabiting urban centres that a single urban dweller could potentially interact with creates an overload of potential stimulation (Milgram 1970). Simmel describes the *blasé* attitude, "an incapacity [...] to react to new sensations with the appropriate energy" during everyday urban life through "the rapidly changing and closely compressed contrasting stimulations of the nerves" (1950). On the other hand, anonymity in urban centres also has positive aspects such as increased tolerance for individuality and eccentricity of urban dwellers belonging to marginal groups (Milgram 1970). All software applications discussed in this chapter have been designed in a way that the shared content is the central focus of the interaction rather than the person behind the content. This aligns with the above-mentioned concept of implicitly giving off information (Goffman 1966) while spending time in public urban places. For example, people demonstrate parts of their identity and belonging through the fashion they choose to wear, or body language and facial impressions might provide insights into the mood of a person. We argue that urban digital design interventions should align to and consider the concept of overstimulation and anonymous urban life in a way that the digital information exchanged represents an additional, given off clue towards collocated people and place. Instead of utilising user profiles or subscriptions and overload potential users with digital information about collocated people, we argue that design interventions should be designed in an anonymous way focussing on content rather than people. Considering the fact that users would be overstimulated with having access to the demographic information of all collocated people around them, it seems obvious that urban design interventions for anonymous collocated people should employ mechanisms to filter out and present information in a way that is most attractive and stimulating for the user. Also considering that in an anonymous public urban place a person only has access to the given off information from collocated people, the question arises why digital design interventions should change that and provide users with a wealth of information possibly resulting in a digital *blasé* attitude. Instead, we argue that the information provided in design interventions for anonymous collocated people should provide a digital lens into the variety of lightweight data describing collocated people and place, ultimately stimulating an *intéressé* attitude towards the variety of lightweight data that collocated people share.

3.3.2 Lightweight Content

Throughout this chapter, the term "lightweight data" has been used. Lightweight data are information that do not breach a user's privacy if they volunteer to share such information. In this study, lightweight data in the form of currently played music, videos, and text comments have been utilised. However, these kinds of lightweight data are only non-privacy-sensitive and lightweight, because the data are anonymised and decoupled from identity. As an example, the web service ijustmadelove.com enables users to share the geographical location of where they just made love. The location in turn is visualised and aggregated on a map without identifying a user. The information shared by ijustmadelove.com could be considered as highly privacy-sensitive information, but the website does not include any identifiable information other than location. As the service received over 240,000 submissions while writing this chapter, it seems that its users do not have strong concerns towards sharing such personal data—albeit in an anonymous way.

Capital Music shares the currently played song choice with collocated people. While music does not seem to be private data, some people might not feel comfortable with granting access for other people to browse their playlist. The concept of *playlistism* describes how people might judge other people on the basis of their personal music taste (Valcheva 2009). Most of the PlaceTagz stickers have been deployed in public toilets enabling urban dwellers to leave a comment on a digital message board. Utilising user profiles and therefore connecting content to an online identity could have resulted in some people feeling uncomfortable due to the context of the interaction: being in a public toilet. YourScreen on the other hand provides private means for video submissions utilising the mobile phone of a user while also avoiding to display any identifying information about the user who submitted the video currently played on the screen.

The proliferation of participatory Web 2.0 services requiring registrations and user profiles has created a trend of collecting data about users for future business model opportunities. We emphasise that software applications designed for anonymous collocated people should avoid this trend and enable interactions without the need to provide personal data. Furthermore, as the applications discussed in this chapter are designed to be used in situ while spending time in an urban public place, requiring a user to fill out a registration form while, for example, waiting for the bus to submit a video to the YourScreen application, would negatively influence the user experience and probably lessen their motivation to use the service.

A potential threat of abusing people and exchanging inappropriate content through anonymising data for lightweight information sharing between unknown collocated people exists. Mechanisms to block and prevent such behaviour should be integrated into design interventions to reduce the probability of abuse. While PlaceTagz did not receive any abusing messages during its real-world deployment, the spam filter employed by the app prevents the use of offensive language in text comments. During the real-world deployment of YourScreen, a small group of users tried to "game" the system by submitting 10-h-long videos, which led us to add a timer function switching to the next video after 5 min of playback. During the simulated real-world evaluation of Capital Music, two user groups could be identified: one group of users who would enjoy arguing about personal music tastes and one user group who would be afraid to receive offensive messages. A simple way to assist the latter group could be to integrate functionality to block and hide users. We believe such mechanisms should provide sufficient protection.

4 Conclusion

In this chapter, we presented three software applications designed to enable collocated people to anonymously augment the urban space with mobile-mediated interactions. Drawing on research from urban sociology, psychology, and urban design, and the findings from our own user studies, we discussed how such interactions can have an impact on the urban experience. Previous research suggests that urban technologies should connect urban dwellers more to physical places and people within (Lentini and Decortis 2010). This chapter demonstrated that mobile technologies and public urban displays in combination with anonymous mobile-mediated interactions and digital augmentations can create new *digital* optional and social activities in space. Additionally, design interventions as discussed in this chapter enable urban dwellers to play with their anonymous digital presentation of self in place. The chapter discussed how—on the individual level—design interventions can help to relieve boredom, and pique and stimulate people's curiosity towards collocated people, place, and the design interventions themselves. This in turn can result in raised awareness towards the environment. Finally, on the technological level, we recommend to design anonymous interactions for collocated people aligning with the anonymity which urban dwellers usually experience in everyday urban life. Furthermore, design interventions should focus on the content of the digital augmentations rather than on people's identity. Our goal was to create a new digital layer of personal information volunteered by users rather than privacy-sensitive information revealing a user's identity.

This chapter provided insights into how anonymous mobile-mediated interactions influence the experience of people in urban public places. While this analysis and discussion provided insights into how the urban experience might be influenced, we hope that these findings can also be used to inform the design of future urban informatics applications enabling interactions between collocated people. For example, designers can consider the questions: how can design interventions enable users to be playful with the digital presentation of self in place, decoupled from the physical presentation of self? How can design interventions pique curiosity towards collocated people and place? How can data be anonymously shared in a meaningful way? The findings presented in this chapter have been discovered through a bottom-up approach: three different applications have been designed with an emphasis on what data can be shared in particular public urban places followed by user experience evaluations. Insights from these three studies led to this synthesis chapter.

Acknowledgements This research was carried out as part of the activities of, and funded by, the Smart Services Cooperative Research Centre (CRC) through the Australian Government's CRC Programme (Department of Innovation, Industry, Science and Research). We would also like to thank all study participants.

References

Anderson, S. P. (2011). *Seductive interaction design: Creating playful, fun and effective user experiences*. Berkeley, CA: New Riders.

Bargas-Avila, J. A., & Hornbæk, K. (2011, May 2–12, 2011). *Old wine in new bottles or novel challenges? A critical analysis of empirical studies of user experience*. Paper presented at the CHI 2011, Vancouver, BC, Canada.

Bassoli, A., Brewer, J., Martin, K., Dourish, P., & Mainwaring, S. (2007). Underground aesthetics: Rethinking urban computing. *Pervasive Computing, IEEE, 6*(3), 39–45.

Bell, G. (2011). *The value of boredom. TEDxSydney*. Retrieved May 14, 2012 from http://tedxtalks.ted.com/video/TEDxSydney-Genevieve-Bell-The-V.

Berlyne, D. E. (1954). A theory of human curiosity. *British Journal of Psychology, 45*(3), 180–191.

Birnbaum, M. G. (2008). *Taking Goffman on a tour of facebook: College Students and the presentation of self in a mediated digital environment*. Tuscon: University of Arizona.

Bull, M. (2012). The end of flanerie. In Ulrick Ekman (Ed.), *Throughout: Art and culture emerging with ubiquitous computing* (pp. 151–162). Cambridge MA: MIT Press. ISBN 9780262017503.

Conrad, P. (1997). It's boring: Notes on the meaning of boredom in everyday life. *Qualitative Sociology, 20*(4), 465–475.

Crawford, A. (2008). Taking social software to the streets: Mobile cocooning and the (an-)erotic city. *Journal of Urban Technology, 15*(3), 79–97.

Farman, J. (2011). *Mobile interface theory: Embodied space and locative media*. London, UK: Routledge. ISBN 978-0415878913.

Fisher, C. D. (1993). Boredom at work: A neglected concept. *Human Relations, 46*(3), 395–417.

Fortunati, L. (2005). Mobile telephone and the presentation of self. In R. Ling & P. E. Pedersen (Eds.), *Mobile communications re-negotiation of the social sphere* (Vol. 31, pp. 203–218). London: Springer.

Foth, M. (2011). *Urban informatics, ubiquitous computing and social media for healthy cities*. Paper presented at the 4th Making Cities Liveable Conference. Noosa, Queensland, Australia. http://eprints.qut.edu.au/42589/

Foth, M., Choi, J. H, & Satchell, C. (2011). Urban informatics. In *Proceedings of the ACM 2011 Conference on Computer Supported Cooperative Work (CSCW '11)* (pp. 1–8). New York, NY, USA: ACM. doi:http://dx.doi.org/10.1145/1958824.1958826.

Foth, M., & Schroeter, R. (2010). Enhancing the experience of public transport users with urban screens and mobile applications. In *Proceedings of the 14th International Academic MindTrek Conference: Envisioning Future Media Environments (MindTrek '10)* (pp. 33–40). New York, NY, USA: ACM. doi:http://dx.doi.org/10.1145/1930488.1930496

Gehl, J. (1987). *Life between buildings*. New York: Van Nostrand Reinhold Company.

Gehl, J. (2010). *Cities for people*. Washington, DC, USA: Island Press.

Goffman, E. (1959). *The presentation of self in everyday life*. Harmondsworth, Middlesex, England: Penguin Press.

Goffman, E. (1966). *Behavior in public places*. New York: The Free Press.

Gopinath, S., & Stanyek, J. (2014). *The oxford handbook of mobile music studies*. Oxford, UK: OUP. ISBN 978-0-19-537572-5.

Gordon, E., & de Souza e Silva, A. (2011). *Net locality: Why location matters in a networked world*. Wiley-Blackwell.

Hassenzahl, M., & Tractinsky, N. (2006). User experience—a research agenda. *Behaviour and Information Technology, 25*(2), 91–97.

Junglas, I. A., & Watson, R. T. (2008). Location-based services. *Communications of the ACM, 51*(3), 65–69.

Kashdan, T. B., Rode, P., & Fincham, F. D. (2004). Curiosity and exploration: Facilitating positive subjective experiences and personal growth opportunities. *Journal of Personality Assessment, 82*(3), 291–305.

Kindberg, T., Chalmers, M., & Paulos, E. (2007). Guest editors' introduction: Urban computing. *Pervasive Computing, IEEE, 6*(3), 18–20.

Law, E. L. C., Roto, V., Hassenzahl, M., Vermeeren, A. P. O. S., & Kort, J. (2009). *Understanding, scoping and defining user experience: A survey approach.* Paper presented at the 27th international conference on Human factors in computing systems, Boston, MA, USA.

Lechner, F. J. (1991). Simmel on social space. *Theory, Culture and Society, 8*(3), 195–201.

Lentini, L., & Decortis, F. (2010). Space and places: When interacting with and in physical space becomes a meaningful experience. *Personal and Ubiquitous Computing, 14*(5), 407–415.

Loewenstein, G. (1994). The psychology of curiosity: A review and reinterpretation. *Psycological Bulletin, 116*(1), 75–98.

Mainwaring, S. D., Anderson, K., & Chang, M. F. (2005). *Living for the global city: Mobile kits, urban interfaces, and ubicomp.* Paper presented at the UbiComp 2005: Ubiquitous Computing.

Marx, G. T. (1999). What's in a name? Some reflections on the sociology of anonymity. *The Information Society, 15*(2), 99–112.

Milgram, S. (1970). The experience of living in cities. *Science, 167*(3924), 1461–1468.

Miller, H. (1995). *The presentation of self in electronic life: Goffman on the Internet.* Paper presented at the Embodied Knowledge and Virtual Space Conference.

Pavlov, I. (1927). *Conditioned reflexes.* Oxford, England: Clarendon Press.

Schau, H. J., & Gilly, M. C. (2003). We are what we post? Self-presentation in personal web space. *Journal of Consumer Research, 30*(3), 385–404.

Seeburger, J. (2012a). No cure for curiosity: linking physical and digital urban layers. In *Proceedings of the 7th Nordic Conference on Human-Computer Interaction: Making Sense Through Design* (NordiCHI '12) (pp. 247–256). New York, NY: ACM. doi:10.1145/2399016.2399054.

Seeburger, J. (2012b). *Influencing the experience of people in urban public places through mobile mediated interactions.* Paper presented at the Proceedings of 3rd International Workshop on Pervasive Collaboration and Social Networking (PerCol 2012), Lugano, Switzerland.

Seeburger, J., Foth, M., & Tjondronegoro, D. (2010). Capital music: personal expression with a public display of song choice. In *Proceedings of the 6th Nordic Conference on Human-Computer Interaction: Extending Boundaries (NordiCHI '10)* (pp. 777–780). New York, NY, USA: ACM. doi:http://doi.acm.org/10.1145/1868914.1869027.

Seeburger, J., Foth, M., & Tjondronegoro, D. (2012a). Designing and evaluating mobile multimedia user experiences in public urban places: Making sense of the field. In D. Tjondronegoro (Ed.), *Mobile multimedia—user and technology perspectives* (pp. 117–132): Croatia: Intech.

Seeburger, J., Foth, M., & Tjondronegoro, D. (2012b). The sound of music: sharing song selections between collocated strangers in public urban places. The sound of music: sharing song selections between collocated strangers in public urban places. In *Proceedings of the 11th International Conference on Mobile and Ubiquitous Multimedia (MUM '12)* (Article 34, 10 pages). New York, NY, USA: ACM. doi:http://dx.doi.org/10.1145/2406367.2406409.

Simmel, G. (1950). The metropolis and mental life. In K. Wolff (Ed.), *The sociology of georg simmel* (pp. 409–424). Glencoe, Illinois: Free Press.

Sutko, D. M., & de Souza e Silva, A. (2011). Location-aware mobile media and urban sociability. *New Media Society, 13*(5), 807–823.

Townsend, A. M. (2000). Life in the real-time city: Mobile telephones and urban metabolism. *Journal of Urban Technology, 7*(2), 85–104.

United Nations. (2008). *World urbanization prospects: The 2007 revision.* New York, NY: Department of Economic and Social Affairs, United Nations Secretariat.

Valcheva, M. (2009). *Playlistm: A means of identity expression and self-representation.* University of Oslo: Unpublished Technical Report.

Wei, R., & Lo, V.-H. (2006). Staying connected while on the move: Cell phone use and social connectedness. *New Media and Society, 8*(1), 53–72.

Narratives of Ambient Play: Camera Phone Practices in Urban Cartographies

Larissa Hjorth

Abstract This chapter explores the unofficial role of camera phone practices in visualising everyday forms of ambient play. I argue that camera phone practices—especially in an age of geo-tagging—are creating their own cartographies of place that overlay the visual with the ambient, social with the geographic, and emotional with the electronic. In other words, camera phone practices evoke the ongoing importance of ambient play and co-presence in mapping a sense of place. Having outlined the notion of performative cartography as part of what has been defined as 'critical cartography', I consider how camera phone practices can be understood through ambient, co-present play. I turn to a site-specific mobile game that deploys Instagram to explore the ways in which cartography can be performed and how that performativity creates new ways for engaging with an everyday place. The game, *keitai mizu* [mobile water]), was made for a Tokyo post-3/11 tsunami and Fukushima disaster context. *Keitai mizu* renders players into investigators by using the camera phone and Twitter as part of the discovery process in uncovering the natural water streams under the urban cartographies.

Keywords Urban cartographies · Mobile phones · Mobile phone cameras · Photography · Ambient play · Mobile games

Within urban spaces, the taking, reflecting and sharing of camera phones are redefining the overlay between spatial, temporal, social and cultural narratives. As mobile phone apps such as Instagram emplace and overlay the electronic with the geographic and the social with the location, they are reshaping the potential

L. Hjorth (✉)
School of Media and Communication, RMIT University, GPO Box 2476, Melbourne VIC 3001, Australia
e-mail: larissa.hjorth@rmit.edu.au

M. Foth et al. (eds.), *Citizen's Right to the Digital City*,
DOI 10.1007/978-981-287-919-6_2

cartographies of the urban (Gordon and de Souza e Silva 2011; Farman 2011; Perkins 2009; Wilmott 2013). Through user-created content (UCC) and locative maps, mobile media are creating new forms of spatial mapping (Farman 2011). This feature has not been lost on many new media artist collectives such as Blast Theory or artists such as Drew Hemmet, who argue for the importance in investigating the relationship between space and cartography (Farman 2011). Artists such as Christian Nold (2009) and theorists such as Nanna Verhoeff (2013) have been key in thinking about cartography as performative. This performative dimension is no more apparent than in the case of camera phone practices in urban spaces. They meditate and re-present. They reframe. They play a powerful role in the experience, representation and performance of the urban.

And yet, few studies on camera phones have connected the potential citizenship and vernacular agency around apps such as Instagram. There is a disconnect between the studies around identity work (Wendt 2014; Walker Rettber 2014), creative practice/art practice (Rieser 2011; Zylinska 2015; Palmer 2014) and its role in reconfiguring how we experience and represent place (Farman 2011) and citizenship. Mobile media play a key role in emergent intimate publics whereby binaries such as online/offline, public/private and work/personal are eschewed. This entanglement can be understood through the notion of ambient play.

By framing mobile art and game interventions in terms of ambient play, this chapter seeks to curate some of the undulating entanglements of mobile media within everyday life. I argue that camera phone practices—especially in an age of geo-tagging where images are encoded with geographic information—are creating their own cartographies of place within the urban as they overlay the visual with the ambient, social with the geographic and emotional with the electronic. In other words, camera phone practices evoke the ongoing importance of ambient play and co-presence in mapping a sense of place. In this chapter, I discuss a site-specific mobile game that deploys Instagram to explore the ways in which cartography can be performed and how that performativity creates new ways for engaging with an everyday place. The game, *keitai mizu* [mobile water], was made for a Tokyo post-3/11 (post-tsunami and Fukushima disaster) context.

The game takes place in an urban park in the busy location of Shibuya. Underneath the park is a series of hidden streams. The aim of *keitai mizu* was to make pedestrians aware of these streams and to reflect differently upon the park space. In order to do so, we hid a series of Japanese and Australian artworks of water creatures around the park. Players had 10 min to search the park and uncover, photograph and share (via Instagram or Twitter) the artwork. However, some of the artworks looked like rubbish to further confuse the relationship between art and the quotidian, game and non-game, and water and non-water spaces. *Keitai mizu* renders players into investigators by using the camera phone and Twitter as part of the discovery process in uncovering the natural water streams under the urban cartographies.

In order to explore the role of camera phones in the re-narrating and intervening of urban spaces, this chapter firstly outlines changing definitions of ambient play and cartography in the light of mobile media and camera phone apps. Then, I turn to the case study of *keitai mizu* to consider how the use of camera phones as part of gameplay can provide new ways to co-presently intervene and represent urban spaces.

1 Locating Ambient Play

Camera phone practices have become an integral part of everyday life. Camera phone apps such as Instagram, while often playing an unofficial role in everyday practices, have seemingly colonised every experience. No experience is too banal or quotidian to photograph and share; rather, as Ilpo Koskinen has argued, camera phone practices are ordered by the logic of the banal and mundane (2007). In their everydayness, they reflect and amplify the rhythm and movements across places, spaces and temporalities. Camera phone practice shapes, and is shaped by, different modes for conceptualising place.

As they move in and out of the background and foreground in urban spaces, camera phone practices contribute to what can be called ambient play. Ambience is often used to describe the effects of sound and music; or in human–computer interaction (HCI), it has played a key role in the understanding of ubiquitous computing (Dourish 2005); as a noun, it specifically refers to a style of music with electronic textures and no consistent beat that is used to create a mood or feeling, but more generally, the term alludes to the diffuse atmosphere of a place. In short, ambience is about the sensorial and affective *texture* of a place. There are many features of gameplay that are ambient—most explicitly the soundtracks that play a pivotal role in developing the mood, genre and emotional clues for the player. Without their soundscapes, many games would fail, and yet, like ambience, the importance of sound is also relatively overlooked in games studies despite its pivotal role in player embodiment.

However, what constitutes ambience within the context of mobile gameplay—especially as it moves across different modes of mediated and co-located presence, and thus different experiences of emplacement—means we need to develop a more robust understanding of ambience (Hjorth and Richardson 2014). In mobile games, the audiovisual ambience is augmented by the haptic, social, networked and locative elements of the game experience. As we have argued in other work, interpreting mobile gaming as ambient play 'contextualizes the game within broader processes of sociality and embodied media practices', and defines play as something that takes place both in and out of games, reflecting broader cultural nuances and phenomena (Hjorth and Richardson 2014: 60). So ambience is not just an aural experience, but also conveys the way games infiltrate our social and emotional lives, afford particular sense perceptions and impact upon our movement through domestic and urban spaces. All of these elements work to diversify our experience of co-presence, of being with others; indeed, co-presence in its various forms and combinations is an important part of the ambient texture of gameplay and is what makes online games so compelling.

While the cultural dimensions of play have been discussed in detail by Sutton-Smith (1997) and in the context of games by Zimmerman and Salen (2004) who draw upon Huizinga (1955) and (Caillois and Barash 2001), the playful within spaces like the urban have a long history that can be linked to historical motifs like the *fláneur* and the 1960s movement *Situational International* (de Souza e Silva and Hjorth 2009). In keeping with these historical and sociocultural dimensions of play, in this chapter I argue that camera phone practice expands upon these earlier phenomenon where representing place involves various forms of co-presence (Farman 2011). As Miguel Sicart argues in *Play Matters* (2014), play is increasingly becoming recognised as an integral part of all facets of life and no longer relegated to compartmentalised areas. Play is integral to being human and has multiple cultural, social, historical and emotional entanglements.

Ambient play suggests a need for nuanced and dynamic readings of mobile media as it moves in and through place. As Paul Dourish argues, 'ambience draws our attention to distractions between focus and periphery… and different ways in which information can be incorporated into an environment' (2005: 25). Drawing upon the work of Brown and Duguid (1994), Dourish argues for an embodied phenomenological approach. Elsewhere, he encourages us to understand information, ambience and intelligence as cultural categories that are contested and re-imagined (Dourish et al. 2005). Ambience, along with interaction, has become key cornerstones in HCI and especially ubiquitous computing and yet must remember that they are culturally specific and informed. For Kjeldskov et al. (2013), the term 'digital urban ambience' is a more nuanced lens for understanding the role of mobile devices mediating urban contexts.

Ambient play brings together the growing importance of accounting for ambience in a Dourish definition as a movement between the focus and periphery (2005) and play as central to urban cartographies. The rising significance of play within everyday quotidian life can be paralleled to shifts in definitions of cartography in what has been called the 'critical cartography' turn. For Chris Perkins (2012), the shift in maps towards the performative (Verhoeff 2012) can be understood as part of a broader ludic shift whereby playing no longer happens *on* maps but *in* maps. Drawing on Raessen's (2006) discussion of the ludification of culture whereby play increasingly performs a key role in everyday life, Perkins argues that cartography has learnt from location-based mobile gaming and the ways in which playful collaboration can lead to new ways of seeing places. As Perkins notes, applications such as desktop mapping and GIS have democratised the tools for cartography and in turn made mapping 'no longer tied to fixed specifications' (2012: 2).

Perkins' work is especially important in the context of bringing together the work around play and critical cartography. Camera phone apps expand upon this earlier wave of democratised cartographic tools in ways which are more playful and ambient. And yet, much of the contextualisation of camera phone practices is linked to debates around the remediation of visual culture and the haunting of the analogue within the digital. For Perkins, the ludic or playful turn in culture limits the historical significance of play within many facets of everyday life. As Perkin argues, 'people have always *played*, and perhaps increasingly *play* with mapping, instead of simply making or using a map for an instrumental task' (2012: 3). Elsewhere, Perkins has argued how the rise of vernacular mapping in tandem with cross-disciplinary approaches (i.e. ethno-methodology, actor–network theory, non-representational theory and phenomenology) provides new ways for rethinking mapping where the playful is intrinsically interwoven within its logic (Perkins 2009; Dodge et al. 2009).

Second-generation camera phone practices—as a practice embedded within the movements of the everyday through geo-tagging—is indicative of the creative and performative aspects of cartography (Verhoeff 2013). In particular, they highlight the shift of camera phone practices from representing first-generation 'networked visuality' (Ito and Okabe 2005; Mørk Petersen 2009; Burgess 2007) to what myself and Sarah Pink have called 'emplaced visuality' (2012). With locative media in the form of geo-tagging creating different overlays between cartographies of the geographic, social, electronic and temporal, we need to conceptualise the role camera phones take in the playful adaptation of maps and sense of place.

Unquestionably, we see how embedded camera phone practices have become for making sense of everyday life and co-present intimacy across various temporalities, spatialities and histories. With the rise of smartphones and apps that make for easy sharing via social media while on the run (as opposed to first generation where most people uploaded via the computer), movement across different intimate co-presents becomes apparent. They also contribute to the shaping of place by overlaying digital maps with co-present socialities. As Jason Farman notes in his important study on mobile interface theory, mobile media rewrite our relationship to cartography and place (2011). Concurrent with the shift in first- to second-generation camera phone, studies have been the movement of LBS into the mainstream.

With LBS added to the overlay particularly through geo-tagging, camera phones partake in a variety of emplacements across electronic, social, spatial and temporal distances and closeness. This means a departure from the dominant 'network' paradigms in visual/media culture and Internet studies that suppress the significance of movement as a mere node, towards a focus on 'emplacement' whereby people, images and technologies are always situated, in movement and part of and constitutive of place. Emplacement that is how place is situated through and within the movements is central in practice and thus how we conceptualise microblogging. While place has always mattered to mobile media, this is amplified in an age of smartphones as the increasingly dominant, if not only, portal for social media.

Through sharing playful pictures of places as part of everyday movements, camera phone practices provide new ways of mapping place beyond just the geographic: They partake in adding social, emotional, psychological and aesthetic dimensions to a sense of place. Camera phone image sharing illustrates the significant role of co-presence sociality in the practice of place as something more than just geographic or physical. Within the movement of the everyday, camera phone practices emplace across *temporal, spatial* and *social* registers. For example, geo-tagging emplaces the moment to the spatial, while retro filters emplace to temporal nostalgia. As Nathan Jurgenson argues, the temporal aesthetics of mobile apps such as Instagram are ordered around a 'nostalgia for the present' (2011). As noted in a previous study with Pink LBS camera phone, practices push beyond being a 'snapshot' whereby time and space are frozen—instead, they create new temporal–spatial visual configurations that are embedded with moving cartographies.

Having married the conversations between critical cartography, camera phone apps and ambient play that are informing urban spaces, this chapter now turns to a site-specific mobile game that deploys Instagram and Twitter to create performative cartographies around water and the climate in Japan. The game, *keitai mizu* [mobile water]), was made for a Tokyo post-3/11 tsunami and Fukushima disaster context. *Keitai mizu* renders players into investigators by using the camera phone and Twitter as part of the discovery process in uncovering the natural water streams hidden under the urban cartographies. The game utilises the ambient play of mobile media to allow players to discover their own water story of the place. The shared Instagram pictures offer interpretative, playful and reiterative ways to understand the gameplay space (a park in Shibuya, Tokyo).

2 *Keitai Mizu*: A Site-Specific Mobile Game

> When the earthquake occurred, I was alone in my room playing a monster hunter PSP (PlayStationPortable) game. Exactly at the time, I was fighting with a monster who makes an earthquake so that I didn't realise that an actual, offline quake had occurred. Only after beating down the monster, I realised something different around me. A fish tank had overflowed and books had fallen down. Initially, I was not really shocked by the earthquake itself, but felt frustration with the aftermath—the power failure, panic buying, nuclear accident and such stuff. During this time, I stayed inside with a friend and continued to play the monster hunter game. But the game was no longer entertaining. ('Toshi', 25 Japanese male)

The quote from 'Toshi' sees him playing a haptic game during the 2011 Tokyo earthquake, tsunami and earthquake disaster known as 3/11. Affective and personal technologies such as social and mobile media make us rethink old psychological models of emotion. In times of trauma, mobile media are increasingly becoming a vehicle for material and immaterial textures and contours of grief. Toshi's immersion within the PSP game was so deep that he mistook the quake vibrations for the monster's movements within his game. In the moments after he realised the horror of the real-life event, all he desperately tried to do was to contact friends and family but to no avail. In the days after 3/11 and as multiple and conflicting news reports emerged across mass and social media, Toshi played the game with a friend to hide from the pain and confusion. Later, it emerged that the national broadcaster, NHK, had deliberately withheld important information about the Fukushima reactor under the instructions of the government. For Toshi—like millions of other Japanese—they shifted their trust towards mobile media such as Twitter and LBS like Foursquare and Instagram to help them not only gain a sense

Fig. 1 *Shibuya* underground streams

of intimate publics but to also have a sense of perpetual co-presence with their family and friends.

Given the important role played by camera phone apps such as Instagram and mobile apps such as Twitter within everyday Japanese urban life, we developed a site-specific mobile game, *keitai mizu*. *Keitai mizu* was motivated by the question—how can we harness Twitter and camera phone apps to make a game that deployed ambient play and co-presence to reflected upon the environment in new ways? It was in a post-3/11 context that the art and climate change group *Spatial Dialogues*[1] began to collaborate with the Japanese Boat People Association to develop different artistic ways to map Tokyo sites in terms of the hidden streams, called *Shibuya: Underground Streams*. Given that Tokyo is made up of numerous little rivers underneath all the trains and roads we wanted to make audiences aware that they are literally perpetually *walking on water*.

Through a series of video, sound, game and sculptural narratives, *Shibuya: Underground Streams* sought to make the general public in Tokyo consider the underground streams constituting much of a hidden Tokyo. In particular, the project focused upon one of the busiest places in the world, Shibuya. By placing a shipping container in a park over the month of June 2013, our study sought to explore the idea of cartographies—water, emotional, social, playful, psychological, historical and geographic. One component involved and boat out of water and lost fisherman in the busy intersection of Shibuya crossing (see Fig. 1).

[1]*Spatial Dialogues* was an Australian Research Council Linkage project.

Japanese and Australian artists were asked to make a series of abstract and representational works of water creatures, which were then placed around the park. The project sought to disrupt dichotomies between art and non-art, water and non-water, game and non-game, seen and hidden, and player and ethnographer. Players had 10 min to hunt for, photograph and share in online various *native-only* water-related creatures and objects that have been placed around the site. They then 'captured' the art with their camera phones and shared it online on Twitter or Instagram. Winners only sent pictures of the native species to the @keitaimizu Twitter account (Fig. 2). The game deployed both old (geo-caching) and new (Twitter and Instagram) media to turn players into ethnographers.

Through playing with ambience, the hidden water streams were brought to the forefront through artworks placed around the park. For many of the participants, they were unaware of the rivers underneath the roads and train tracks. The fact that in many cases, it was hard to distinguish between the art and rubbish in the park meant players were constantly having to analyse all the everyday objects as potential artworks. The slippage between the intentional objects of the game and what was already there further defused the relationship between the art and non-art, and game and non-game space. The artwork was ambient, moving in and out of the background while setting the mood.

Co-presence was a key feature of the game. As players went around photographing and sharing images online, co-present friends (i.e. friends geographically absent but electronically present) began to collaborate and query the pictures taken. Far from eroding a relationship between absence and presence, *keitai mizu* sought to highlight the importance of ambient co-presence in the construction and experience of place. Some co-present friends were co-opted into the game space by trying to help their friends solve the game by guessing what was the art

Fig. 2 *Keitai mizu* (mobile water) game

and what was the native to Tokyo water creatures. The game space intentionally blurred online and offline spaces with Instagram and Twitter enabling co-present friends to share the experiences and images. Through the process of gameplay, participants became more mindful of the local water species as well as reflective upon the fact that the city is made up of numerous little rivers underneath all the trains and roads. We wanted to make audiences aware of the hidden water cartographies and how mobile media apps demonstrate how maps are performative—we shape maps as they shape us.

Keitai mizu attempted to challenge boundaries between official and unofficial game spaces by blurring them with different modes of play (Fig. 3). In particular, camera phone practices partake in new haptic visualities that bring emotional and social dimensions of place and play to the official gameplay space and drive the motivation for use. By deploying camera phone practices as part of the mobile game, players can develop the melodramatic elements—the affective and emotional dimensions—to engage friends into the play of being mobile.

Through the playful use of Instagram geo-tagging whereby numerous images of artefacts were assembled upon the Website, players were able to see other players' choices (i.e. what they thought were the native animals) and their location through geo-tagging. This created a sense of emplacement but also displacement as other players searched for some art objects that were either mistaken for

Fig. 3 *Keitai mizu* players

rubbish in the park or too small to see (some artworks, such as Yasuko Toyoshima, were semi-transparent creatures only 5 cm long). The *Spatial Dialogues* Website became a series of emplaced visualities of the park through each of the players' interpretations. The mapping of the park and its underground streams became a series of Instagram clues.

The playing with emplaced visuality was an important component of the game.

As argued elsewhere with Pink (2012), we need a non-representational approach to understanding camera phone photography building on the work of Tim Ingold, Nigel Thrift, Hayden Lorimer and Tim Cresswell. Expanding upon Pink's understanding of photography as produced and consumed in movement and as emplaced within the flow of everyday life, we argued that camera phone photography has little to do with being 'connected' in a 'network'. They are more related to actual, imagined, material, sensory and digital experiences that are affective, embodied and social. In contrast to what he sees as the tendency of network theory to construct a separatedness of things, we called for an understanding of camera phone practices that understand the environment as made up of a 'meshwork of lines', or 'texture' (2014). Rather than camera phones being about networked visuality as in the first generation of studies, the notion of emplaced visuality sees the images as embedded within, and through, the movements of the camera phone photographers with their geo-tagging apps adding new dimensions to spatial and temporal movements.

Fig. 4 Co-present and ambient play examples of *keitai mizu*

Part of the enjoyment of the project was not only the entanglements between the methods and its transmission but also how the project lived on in different ways that saw the participants taking the key role. For example, when one student group came through to play, one of the other students took it on herself to document their experiences and responses and turn it into a short film which she then uploaded onto vimeo. This video was one of the few artefacts of transmission left after the ephemeral work had ceased. Moreover, traces of the play could be found in participants' twitter accounts, creating new nodes for co-present entanglement (Fig. 4).

3 Conclusion: Picturing the Urban

In this chapter, I have proposed an understanding of camera phone practices as an important part of how we experience, represent and perform in and through urban spaces. As an omnipresent and yet ambient part of everyday play, mobile media provide new ways in which we can conceptualise cartographies and rethink the role of the citizen in the urban. They provide a lens in which the public and intimate can be interwoven in new ways—what can be called 'intimate publics' (Hjorth and Arnold 2013). The idea that forms of intimacy might be generated in contexts that are simultaneously public is not new. Writing before social media had become an integral part of everyday worlds, Berlant observed that intimacy has taken on new geographies and forms of mobility, most notably as a kind of 'publicness' (1998, 281). However, in a digital–material environment, intimate relations are not simply performed in pairs or in bounded groups. Rather, they traverse the online and offline in that they are performed in physical public worlds, but also in electronic privacy (e.g. when someone privately sends a camera phone image of themselves in a café to a friend), and in an electronic public that is geographically private (e.g. when we read personal messages posted to us in a publicly facing Facebook page or on Twitter, while in the private space of our homes). This informs how we think and practice citizenship.

Camera phone practices are an essential part of the contemporary citizen's tool kit. Camera phone practices also provide us with ways in which to think about Warner's 'counter publics' (2002) whereby the citizen is involved in often quotidian and unofficial place-making exercises that become an embodied part of how they experience the urban. In recent years, social media and online platforms accessed via mobile technologies have become an integral part of digital activism (Postill 2011). Mobile media has been a key in the increasingly active role of citizens in their making and remaking of the urban. Locations such as Tokyo are indicative of this phenomenon (Slater 2011; Slater et al. 2012). Mobile media afford us with new ways to play not just *with* maps but also *in* maps (Perkins 2012). Through the case study of *keitai mizu*, we see how camera phone apps can be re-deployed in a form of geo-caching to rethink and reflect upon the space of the urban. Moreover, their ability to entangle ambient play with intimate co-presence across different temporality and spatialities means that they can expand upon how we conceptualise the urban.

In exploring camera phone practices in relation to a place-making mobile game, I have sought to provide poetic ways in which players can become investigators in understanding their everyday environments in new ways. As I have argued, camera phone practices contribute to the various performative cartography cultures emerging in and around the urban. They provide more than a thousand words to understand the urban today.

References

Brown, J. S., & Duguid, P. (1994). Borderline issues: Social and material aspects of design. *Human-Computer Interaction, 9*, 3–36.

Burgess, J. (2007). *Vernacular creativity and new media* (Doctoral dissertation). Accessed September 30, 2012. http://eprints.qut.edu.au/16378/.

Caillois, R., & Barash, M. (2001). *Man, play, and games*. Urbana: University of Illinois Press.

de Souza e Silva, A., & Hjorth, L. (2009). Urban spaces as playful spaces: A historical approach to mobile urban games. *Simulation and Gaming*, *40*(5), 602–625.

Dourish, P. (2005). The culture of information: Ubiquitous computing and representations of reality. In *Designing ubiquitous information environments: socio-technical issues and challenges ifip—the international federation for information processing* (Vol. 185, pp. 23–26). New York: Springer.

Dourish, P., Brewer, J., & Bell, G. (2005). Information as a cultural category. *Interactions: Ambient intelligence, 12*(4), 31–33.

Farman, J. (2011). *Mobile interface theory*. London: Routledge.

Gordon, E., & de Souza e Silva, A. (2011). *Net locality: Why location matters in a networked world*. Chichester, UK: Wiley-Blackwell.

Hjorth, L., & Arnold, M. (2013). *Online@AsiaPacific*. London: Routledge.

Hjorth, L., & Richardson, I. (2014). *Gaming in locative, social and mobile media*. London: Palgrave.

Huizinga, J. (1955). *Homo ludens*. Boston: Beacon Press.

Ito, M., & Okabe, D. (2005). *Intimate visual co-presence*. Paper presented at *Ubicomp*, Takanawa Prince Hotel, Tokyo, Japan, September 11–14. Accessed 28 June 2006. http://www.itofisher.com/mito/.

Jurgenson, N. (2011). The Faux-Vintage photo: Full essay (Parts I, II, and III)', *The society pages,* May 14. http://thesocietypages.org/cyborgology/2011/05/14/the-faux-vintage-photo-full-essay-parts-i-ii-and-iii/.

Keitai mizu (2013). Available at: http://spatialdialogues.net/tokyo/keitaimizu/, http://spatialdialogues.net/tokyo/live-in-tokyo/audience-response-to-shibuya-underground-streams/.

Kitchin, R., Perkins, C., & Dodge, M. (2009). *Rethinking Maps*. London: Routledge.

Kjeldskov, J., Skov, M. B., Nielsen, G., Vestergaard, M., & Thorup, S. (2013). Digital urban ambience: mediating context on mobile devices in the city. *Pervasive and Mobile Computing*. *9*(5), 738–749.

Koskinen, I. (2007). Managing banality in mobile multimedia. In R. Pertierra (Ed.), *The social construction and usage of communication technologies: European and Asian experiences* (pp. 48–60). Singapore: Singapore University Press.

Mørk Petersen, S. (2009). Common banality: The affective character of photo sharing, everyday life and produsage cultures (doctoral dissertation), ITU Copenhagen.

Nold, C. (2009). *Emotional cartography: Technologies of the self.* http://emotionalcartography.net/. Accessed April 1, 2013.

Palmer, D. (2014). Mobile media photography. In G. Goggin & L. Hjorth (Eds.), *The routledge companion of mobile media* (pp. 245–255). New: Routledge.

Perkins, C. (2009). Playing with maps. In M. Dodge, R. Kitchin & C. Perkins (Eds.), *Rethinking maps* (pp. 167–188). Routledge: London.

Perkins, C. (2012). Playful mapping: The potential of a ludic approach, international cartographic association conference. URL: http://icaci.org/files/documents/ICC_proceedings/ICC2013/_extendedAbstract/121_proceeding.pdf.

Pink, S., & Hjorth, L. (2012). Emplaced cartographies: Reconceptualising camera phone practices in an age of locative media. *Media International Australia, 145*, 145–156.

Postill, J. (2011). *Localizing the Internet*. Oxford: Berghahn.

Raessens, J. (2006). Playful identities, or the ludification of culture. *Games and Culture, 1*(1), 52–57.

Rieser, M. (2011) *The mobile audiences: Media art and mobile technologies*. USA: ICI Global.

Sicart, M. (2014). *Play matters*. Cambridge, Mass: MIT Press.

Slater, D. H. (2011). Fukushima women against nuclear power: Finding a voice from Tohoku. *The Asia-Pacific Journal: Japan Focus*. Available at: http://japanfocus.org/events/view/117. Accessed February 4, 2015.

Slater, D. H., Nishimura, K., & Kindstrand, L. (2012). Social media, information and political activism in Japan's 3.11 crisis. *Japan Focus*. Available at: http://japanfocus.org/-Nishimura-Keiko/3762.

Sutton-Smith, B. (1997). *The ambiguity of play*. London: Routledge.

Verhoeff, N. (2012). Navigating screenspace: Towards performative cartography. In P. Snickars & P. Vonderau (Eds.), *Moving data: The iPhone and my media*. New York: Columbia University Press.

Verhoeff, N. (2013). *Mobile screens: The visual regime of navigation*. Amsterdam: Amsterdam University Press.

Walker Rettber, J. (2014). *Seeing ourselves through technology*. London: Palgrave Pivot.

Warner, Michael. (2002). Publics and counterpublics. *Public Culture, 14*(1), 49–90.

Wendt, B. (2014). *The allure of the selfie: Instagram and the new self portrait*. Amsterdam: Institute of Networked Cultures.

Wilmott, C. (2013). Cartographic City: Mobile mapping as a contemporary urban practice. *Refractory*. http://refractory.unimelb.edu.au/2012/12/28/wilmott/.

Zimmerman, E., & Salen, K. (2004). *Rules of play: Game design fundamentals*. Cambridge, MA: MIT Press.

Zylinska, J. (2015). *Photomediations*. UK: Open Press.

The Role of Digital Screens in Urban Life: New Opportunities for Placemaking

Martin Tomitsch, Ian McArthur, M. Hank Haeusler and Marcus Foth

Abstract Large digital screens are becoming prevalent across today's cities dispersing into everyday urban spaces, such as public squares and cultural precincts. Examples, such as Federation Square, demonstrate the opportunities for using digital screens to create a sense of place and to add long-term social, cultural and economic value for citizens, who live and work in those precincts. However, the challenge of implementing digital screens in new urban developments is to ensure they respond appropriately to the physical and socio-cultural environment in which they are placed. Considering the increasing rate at which digital screens are being embedded into public spaces, it is surprising that the programs running on these screens still seem to be stuck in the cinematic model. The availability of advanced networking and interaction technologies offer opportunities for information access that goes beyond free-to-air television and advertising. This chapter revisits the history and current state of digital screens in urban life and discusses a

M. Tomitsch (✉)
Design Lab, Faculty of Architecture, Design and Planning,
The University of Sydney, Sydney, Australia
e-mail: martin.tomitsch@sydney.edu.au

I. McArthur
UNSW Art and Design, The University of New South Wales, Sydney, Australia
e-mail: ian.mcarthur@unsw.edu.au

M.H. Haeusler
UNSW Built Environment, The University of New South Wales, Sydney, Australia
e-mail: m.haeusler@unsw.edu.au

M. Foth
Urban Informatics Research Lab, School of Design,
Queensland University of Technology, Brisbane, Australia
e-mail: m.foth@qut.edu.au

M. Foth et al. (eds.), *Citizen's Right to the Digital City*,
DOI 10.1007/978-981-287-919-6_3

series of research studies that involve digital screens as interface between citizens and the city. Instead of focusing on technological concerns, the chapter presents a holistic analysis of these studies, with the aim to move towards a more comprehensive understanding of the socio-cultural potential of this new media platform, how the digital content is linked with the spatial quality of the physical space, as well as the place and role of digital screens within the smart city movement.

Keywords Placemaking · Urban screens · Large screens · Digital screens · Public displays · Media architecture · Urban interaction design

1 Introduction

Large digital screens have become prevalent across today's cities dispersing into everyday urban spaces, such as public squares and cultural precincts (Foth et al. 2013a). Examples, such as Federation Square in Melbourne, Australia, demonstrate the opportunities for successfully using digital screens to create a sense of place (Frenchman and Rojas 2006) and to add long-term social, cultural and economic value for citizens, who live, work and socialise in those precincts (Fatah gen Schieck 2009).

However, the challenge of implementing digital screens in new urban developments is to ensure they respond and adjust appropriately to the changing physical and socio-cultural environment in which they are placed (Vande Moere and Wouters 2012; Ylipulli et al. 2014). Considering the increasing rate at which digital screens are being embedded into public spaces, it is surprising that the applications running on these screens still seem to be following the principles laid out in the cinema era (Foth et al. 2013a). The availability of advanced networking and interaction technologies offer opportunities for information access that goes beyond free-to-air television, wayfinding and advertising.

This chapter revisits the history and current state of digital screens in urban life and provides a review of research studies that involve digital screens as interface between citizens and the 'digital city'. Instead of focusing on technological concerns, the chapter aims to promote a more comprehensive consideration of the potential of this new media platform for placemaking by understanding digital screens as artefacts embedded within a cultural environment and assessing their potential for civic participation.

The chapter first discusses urban screens and media façades and their emergence in public space, before assessing their role as cultural artefacts in urban media environments. It then suggests ways of using urban screens for placemaking based on a discussion of previous research studying digital screens as platform for civic participation and other forms of passer-by engagement. The chapter ends with a discussion of urban media environments in the context of the smart city movement and directions for future research in this area.

2 Defining Digital Screens in Public Space

In academia, there are two common terms er than creating a large scfor describing media displays in urban spaces that are closely related to each other: urban screens and media façades (Tscherteu and Tomitsch 2011; Haeusler et al. 2012). The fundamental difference between urban screens and media façades lies in the relationship between the screen element, the architecture, and the urban setting. Urban screens typically take the form of mid to large-scale screens that can either be freestanding (e.g., at the centre of a plaza) or attached to a building façade (Fig. 1). If an urban screen is attached to a building, the screen and the building façade remain two separated layers both in terms of the technical installation as well as in the way the two elements communicate their function. If urban screens appear as freestanding, independent architectonic elements, they take on the single purpose to communicate media content (McQuire et al. 2009).

Media façades feature a closer integration of the screen and the building layers, if not a complete integration into a new hybrid structure (Haeusler 2009; Kronhagel 2010). The 14th Venice Architecture Biennale—on the theme of 'Fundamentals' curated by Rem Koolhaas in 2014—lists media façades as one of 'the twelve ecologies of the envelope' (kinds of façade that have been influential during the past century), and consequently acknowledges media façades

Fig. 1 Urban screen at a shopping centre in Singapore

as an architectural element, similar to a double façade or a precast concrete façade (Koolhaus 2014). Media façades consist of integrated light sources or kinetic elements and network infrastructures for power and data distribution. Although from a technical perspective image production still takes place on designated parts of the façade, the visual imagery is designed to be perceived as an integrated part of the built architecture (Fig. 2). This interaction between the building and the audience has been described as 'media architecture' (Tscherteu and Tomitsch 2011). However, the term is more recently used to more broadly describe the integration of any form of media with the built environment. Media architecture provides a framework for the "design of physical spaces at architectural scale incorporating materials with dynamic properties that allow for dynamic, reactive or interactive behaviour" (Brynskov et al. 2013). Media architecture interfaces also include media installations that work with the depth of space, in which case it is no longer possible to speak of a screen or a façade (Tscherteu and Tomitsch 2011).

This distinction between urban screens and media facades is important, as the level of embeddedness of a digital screen into the built environment has an effect on its value and use for placemaking. The level of embeddedness here consists of two dimensions, the structural embeddedness and the conceptual embeddedness. Media façades are inherently more embedded structurally into the built environment as they are designed as part of the building façade. They are therefore perceived to be part of the physical architecture of the city (Haeusler 2009), whereas urban screens and other forms of public displays are perceived to belong to the same family as either smartphones or billboards (Ylipulli et al. 2014). The content typically displayed on and associated with urban screens is likely to have shaped this perception, but equally, content designed for a particular media platform should consider how the platform is perceived in a socio-cultural context (Aurigi 2013). It is crucial that designers create a "healthy balance between the place, the type of

Fig. 2 Media façade at the Ryerson school of image and the Ryerson image centre

content, and the intended impact on the space and the public" (Fatah gen Schieck 2009).

3 Digital Screens as Architectural Building Component

The idea of embedding information into the built environment has been around as long as the written language (Weiser 1991). But the beginning of the history of media façades can be traced back to when architects and designers considered for the first time the application of a TV screen as a building component on an external wall. The first steps were taken in the 1930s with the competition or concept entry of Oskar Nitschke's 'Maison de publicité' in Paris, designed between 1934 and 1936. Cedric Price and Joan Underwood's project development for the 'Fun Palace' in 1961 also incorporated some of the first ideas of embedding large screens into the building fabric. Although neither of the two projects were built at the time and have primarily been idea concepts, they have strongly influenced subsequent media façade projects.

Nathan Silver in his book 'The Making of Beaubourg' (1994) lists these early concepts as precursors of Richard Roger and Renzo Piano's successful media façade competition entry for the Centre Pompidou in Paris in 1971. Obviously, the technological opportunities at the time differed from what is possible today and thus the envisioned media façade project for the Centre Pompidou was also never realised. Still, one could argue, based on a reading of Nathan Silver's book that the Centre Pompidou was very close to realising a media façade. Alan Stanton, the person in charge for the Roger/Piano partnership, experimented with various options ranging from fluorescent lights to projection technology and only the very tight timeframe and the financial pressure stopped the realisation in the end.

Media façades famously featured in the 1982 movie Blade Runner directed by Ridley Scott, which showed personal transport vehicles flying past large-scale screens displaying colourful advertising against a dark dystopian cityscape. Today, the 2001 project Blinkenlights is considered to be one of the first media façades that was actually realised (Haeusler 2009), albeit for a temporary period only. The project used standard lamps that were arranged behind the windows of the top eight floors of a high-rise building, creating an 18 by 8 pixel display. The project differs from contemporary media façades, which use light and pixels as building material, with the aim of emphasising certain aspects of the architecture rather than creating a large screen (Haeusler et al. 2012). However, Blinkenlights is noteworthy in the context of this chapter, as it provided a participatory platform for citizens. People were able to generate their own 18 by 8 pixel video message that was subsequently scheduled to play back on the building façade. This application of urban media goes beyond pure advertising purposes, as featured in Blade Runner, and contributes to placemaking in an urban environment by connecting people with their physical environment and with each other.

4 From Digital Signage to Urban Interface

With advancements in the production of LED technologies, commercial media façades offer ever-increasing pixel density, making them suitable for displaying high-resolution content. However, following the above definition of media façades, such high resolution screens may be considered as urban screens if their primary purpose is the display of media content rather than augmenting the architectural topology of a building.

The availability of cheap LCD and LED/OLED technology has given further rise to the proliferation of urban screens, as well as screens in retail outlets and cafés—often used only to display muted, free-to-air TV (Foth et al. 2013a). A recent market analysis suggested that the average global consumer was exposed to digital signage, or 'digital out-of-home' (DOOH) media, for 14 min per week in 2013, representing a 75 % increase since 2007 (PQ Media, 2014). The annual growth rate of DOOH exposure is forecasted to be 8.6 %. However, the majority of digital public screens are still used for advertising and remain far from fulfilling the vision of providing a useful "gateway that connects the physical city and the digital city" (Foth et al. 2013a). Yet, the investment into DOOH media for advertising demonstrates the significance and rapid growth of this new technology. For example, 7-Eleven operates a network of 12,400 screens through which they reach over 200 million viewers monthly, making them one of the largest broadcast TV networks in the U.S.

The increasing availability of sensor technologies theoretically provides more opportunities for equipping digital screens with interactive capabilities, for example through touch (Jacucci et al. 2010; Ojala et al. 2012) or gestures (Ackad et al. 2013; Hespanhol et al. 2015). Adding interactive capabilities makes urban screens a more interesting proposition from an interaction design perspective and opens them up as a participatory community platform (Memarovic et al. 2015). Providing interactivity at the display increases their value as people can walk up and retrieve or contribute specific information thereby extending the use of public screens beyond advertising and entertainment. However, there are a number of challenges associated with creating interactive digital displays that create barriers to the practical realisation of adding interactivity to urban screens:

- People are often not aware of the interactive capabilities of a display (Ojala et al. 2012), or might not notice the display at all (Müller et al. 2009).
- New forms of interaction are required since there is typically no mouse or keyboard available. Touch input, which has become the standard input mechanism on smartphones, is in many cases not practical due to distance between user and display and the form factors of the display (Ackad et al. 2014).
- Interactions at public displays are typically of opportunistic nature (Valkanova et al. 2013) and for short periods. If a passer-by does not see an immediate benefit that they can gain from engaging with the display they will quickly decide to move on. This effect may also lead to passers-by appropriating the interactive

display for purposes other than what it was intended for, such as engaging in game play (Tomitsch et al. 2014).

- Interactive displays in public spaces may need to cater for multi-user interaction (Peltonen et al. 2008; Boring et al. 2011). This raises additional challenges including the need for essential information such as service departure times, or general wayfinding information to remain visible at all times—even when someone is interacting with the display.
- The design of the interactive display needs to take into account the physical and social context (Vande Moere and Wouters 2012; Akpan et al. 2013). For example, the design of an interactive display in a busy city centre at a major transport hub is likely to be vastly different from an interactive display designed for a suburban location.

5 Urban Media Environments as Cultural Spaces

When deploying digital technologies such as urban screens within a public space it is therefore important to acknowledge the socio-cultural reality of that space (Ylipulli et al. 2014). Making best use of the opportunities of digital screens for placemaking means it is necessary to understand the social and cultural practices inherent to users of the place. Schiffer and Miller (1999) suggest that to learn about a culture, one must understand the interactions between cultural setting (as a place that indicates certain types of social behaviour), the artefacts (and how they are used), and people teaching others a social background and social behaviour (cited in Champion and Dave 2002).

Urban screens and media façades can be seen as a type of artefact that contributes to the cultural character of an environment (Tscherteu and Tomitsch 2011). Tscherteu and Tomitsch propose the term 'urban media environment' to reflect these artefacts' impact on public space as well as the cultural setting in which they are embedded. Urban media environments consist of: *media artefacts*, such as urban screens and media façades, but also personal mobile devices that are either integrated in a local media architecture or that are simply used for personal mobile communication; *other artefacts* like buildings, cars, traffic lights, monuments, urban furniture, and so on; and *specific modes of interaction* as the artefacts invite people to behave in a certain way. While artefacts determine modes of interaction in an environment, they are at the same time themselves the result of, or the reaction to, a sequence of interactions that took place in this environment. For instance, department stores are often built at a specific location as a response to a long tradition of trading in that space, or the installation of traffic lights is the consequence of accidents that happened in comparable urban situations. Similarly, the installation of urban screens at a specific location is typically determined by the location's character, for example using an opportunity where traffic slows down or stops, allowing drivers to watch the content on the screen.

The notion of urban media environments suggests a shift in emphasis from the 'hardware' to the more complex web of artefacts and interactions in a specific space, while also considering the history of the space. This highlights the fact that objects in public space do not exist on their own, but are the result of a co-evolution of artefacts and the daily interactions of city dwellers. This interdependency has a temporal dimension. Artefacts and interactions continuously produce the next generation of artefacts and interactions. In that sense, urban media environments can be considered to be 'cultural spaces' (Tscherteu and Tomitsch 2011). 'Culture' here refers to the historical (or evolutionary) dimension of urban life, which is composed of the inhabitants and their behaviours, buildings, technical developments as well as reflections and theoretical and political concepts that form urban life and its development over time.

Strategies for designing urban media environments need to reflect the specific challenges of public space. For instance, public space has a much older history than that of digital cultural spaces like the Internet. Thus, strategies that focus only on designing digital spaces remain of comparatively limited value and it is therefore necessary to consider the physical as well as the social domains of the built environment (Dourish 2006). When compared to other digital products, like websites, mobile apps, and smartphones, large-scale public installations such as urban screens and media façades offer less opportunity for iterative development and refinement. Such projects therefore have to be approached mindfully from a cultural perspective. This means that it is necessary to look beyond the design of a specific single artefact such as an interactive installation to instead consider the effect on the entire urban environment. Design strategies for urban media environments must include methods of assessing, planning and evaluating the long-term effects of a media intervention in a specific urban environment. Tscherteu and Tomitsch (2011) propose the following steps towards implementing a successful strategy for designing urban media environments:

1. It is necessary to understand public space as an environment that consists of (media) artefacts and social interaction. Interactions can be as simple as crossing a square, taking the dog out, or stopping at a traffic light. They can be of ephemeral and invisible nature, or involve visible artefacts, such as cars, traffic lights, buildings and large displays.
2. Artefacts and interaction constitute each other. Specific environments invite people to behave in specific ways: A park entices people to walk slower compared to a noisy and busy street. There are different modes of perception when driving a car versus walking down an alley. When additional artefacts such as large screens are introduced to an urban environment, it is necessary to understand how they influence the holistic system, for instance by comparing the behaviour of city dwellers before and after a media intervention.
3. The introduction of new media artefacts into an existing environment often creates a cultural conflict due to the fact that the display and/or content were produced in another environment. Content displayed on urban screens is often targeted at a mode of perception that belongs to a different cultural

environment such as watching TV at home. It is necessary to understand the cultural shift between environments by analysing the target space and developing an appropriate form of communication.
4. The introduction of media interventions in public space requires several design and feedback cycles as well as the adoption of a long-term perspective. This imposes the additional challenge for city governance, planning and administration to consider the long-term impact of large-scale displays and other media interventions on urban life.
5. A long-term development can only take place when urban media environments are recognised as a critical challenge for society. It is therefore desirable to look beyond the technical aspects of deploying media artefacts in urban environments. Ultimately, there has to be a harmonious balance between citizens' basic needs in public space and municipal or economic interests. The design of media in urban life should be considered a basic need that in the future will be as self-evident as the need for appropriate transport systems or waste management.

6 From Information Displays to Participatory Civic Platforms

Many contemporary urban screen deployments fail to meet the aspirations and expectations of private and public stakeholders (Vande Moere and Wouters 2012). Despite significant investments by governments and public authorities, these screens are not always used to their full potential for establishing and supporting the local culture and community. In many cases, the issue is that there is no clear understanding of what is required to succeed and of the challenges that come with creating and curating content for a large public screen 24 h a day, 365 days a year. Many existing screens languish as highly visible public demonstrations of the failure of public investments. A stark example is the BBC's withdrawal from its extensive *Big Screens* program, leaving local governments across the UK with the responsibility of managing urban screens. The lack of information and policies for setting up and programming urban screens, leads screen operators to employ large-scale advertising strategies instead of developing more sustainable, curated strategies (Fatah gen Schieck et al. 2009). This diminishes the role that urban screens could otherwise play to contribute to placemaking in urban environments.

One of the challenges with implementing urban screens in new city developments is that they need to respond appropriately to both the physical *and* sociocultural environment in which they are placed (Vande Moere and Wouters 2012; Akpan et al. 2013). Previous attempts to implement the Federation Square model have had only limited success, largely due to the fact that there was a lack of interdisciplinary consultation during the implementation process. It is not uncommon in such developments that the architect or builder hand over the screen to the precinct managers without providing long-term strategies for how the screen can be used to successfully engage local citizens.

As 'big data' (Mayer-Schönberger and Cukier 2013), urban informatics (Foth 2009), and media architecture (Brynskov et al. 2013) attract increasing attention globally, researchers, governments, urban planners, curators and creatives are now questioning how they can support sustainable urban development, civic innovation and economic productivity through community engagement and social participation. Within this unfolding context, local governments, the cultural industries, design and business consultants, developers and urban planners face the complex challenge of creating a sense of place that adds real long-term value for the citizens who live and work in new urban developments.

The organisations behind the urban screens at New York Times Square and Federation Square have successfully demonstrated the potential of urban screens for supporting culture and engaging the local community (Haeusler et al. 2012; p. 145). For example, New York Times Square's art program *Midnight Moment* is a highly popular event that displays synchronised creative content across an array of screens. The Federation Square screen is not only used for broadcasting important cultural events which bring the local community together, but has also been used to adopt community engagement mechanisms such as allowing the public to send messages to the screen via Twitter and text messages (Schroeter et al. 2012).

7 Empowering Citizens Through Digital Screens

It has been argued that people can no longer be understood as passive users or consumers of products and services since they are increasingly able and empowered to also be producers (Foth et al. 2013b; Bruns 2008). Foth et al. (2013b) list three forms of contemporary citizen activism, the first being non-government organisations focusing on broad societal issues, the second referring to organisations that predominantly use online campaigning with some level of user participation (such as clicking a 'Like' button or sending a petition email), and the third form—which is most interesting in the context of this chapter—relating to in situ activism that combines physical and digital forms of engagement. In this third form, digital media is often used for organising ad hoc campaigns, such as Park(ing) Day or Dîner en Blanc. In some cases however, the in situ intervention itself utilises digital media such as the projection of environmental messages onto a wall (Fritsch and Brynskov 2011).

It is indeed possible, and has been suggested, that urban screens and media façades can provide a canvas for citizens to change the visual appearance of the cityscape (Hespanhol et al. 2014). This can alter the emotional relationship citizens form with the city and its building blocks (Tomitsch 2014). Such interventions are akin to popular activism strategies, such as urban gardening and guerrilla knitting, although the digital alteration of public space is of a more ephemeral nature and allows potentially anyone to contribute without requiring special skills or risking legal action.

Another form of civic participation that does not directly alter the visual appearance of the city, but instead gives citizens a voice in longer-term urban planning decisions, is the process of engaging citizens through a consultation process. Traditionally, this consultation is done via face-to-face workshops or online forms, and only reaches certain demographics of the population. As a consequence, the opinions of community members classified as 'hard to reach' may not be reflected in the community consultation process (Sarkissian et al. 2009; Schroeter 2012).

Digital screens provide many new opportunities for creating a direct dialogue between public administration and citizens (Gianluca et al. 2013). This dialogue is a critical aspect in urban planning, ensuring that precincts retain their unique character and qualities (Fredericks and Foth 2013), to allow deliberation of public opinion, and to support discussion of alternative perspectives (Valkanova et al. 2014). The advantage of using situated public screens compared to traditional methods of civic participation such as face-to-face meetings and online surveys can be leveraged when they provide additional channels of engagement to help reach a broader section of the local community, do not require dedicated time and effort, and are embedded within the socio-cultural context (Schroeter et al. 2012; Valkanova et al. 2014). A number of previous studies have investigated public screens for civic engagement, ranging from low-cost interactive posters (Vlachokyriakos et al. 2014) to gesture-based large projection displays (Valkanova et al. 2014) and media façades (Behrens et al. 2014; Korsgaard and Brynskov 2014).

A common issue observed in field trials of such urban interfaces is a lack of participation from the public (Gianluca et al. 2013; Valkanova et al. 2014). People do not yet typically expect public displays to be interactive (Ojala et al. 2012) and either do not notice the interfaces or worry about embarrassing themselves in public (Behrens et al. 2014). These effects present barriers to civic participation of this nature. However, previous field trials show that once people have overcome those barriers, they express feelings of empowerment and connectedness with the local government (Taylor et al. 2012).

Hespanhol et al. (2015) list the following key factors for successfully deploying situated engagement mechanisms: (1) raising awareness about the opportunity to engage in a civic dialogue among passers-by; (2) motivating people to participate in the dialogue; (3) creating a balance between visibility of the interface and privacy during the engagement process, and; (4) provision of effective and timely feedback on the participation process. They investigated these key challenges by studying a series of civic interfaces deployed on and around an urban screen. The urban screen was situated in a public space located in the heart of an urban cultural zone featuring a theatre, a concert hall, and a number of shops and restaurants (Fig. 3). The urban screen is positioned on a terrace about 6 metres above the public space. The lack of proximity to users in the area makes it challenging to implement direct interaction mechanisms. The study found that by displaying an interface on the urban screen that showed a live feed of the interaction space and allowing passers-by to interact through gestures (Fig. 4, right), led to higher participation rates when compared to collecting input from passers-by through a tablet

Fig. 3 The concourse with the urban screen (*top left*) installed on a terrace about 6 metres above the open public space

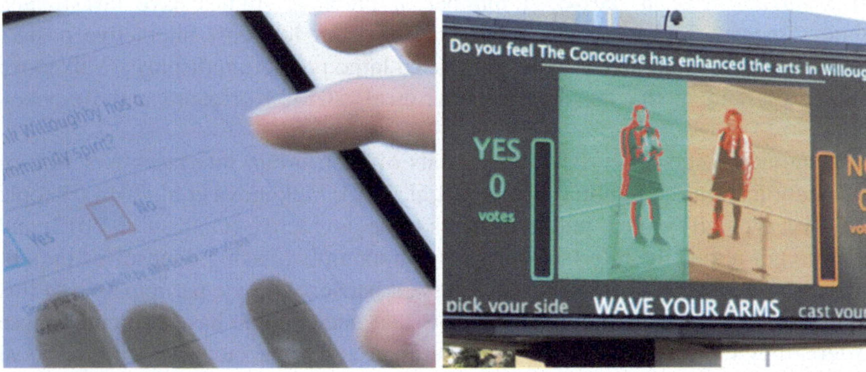

Fig. 4 The tablet voting application (*left*) and the full-body voting application running on the urban screen (*right*)

app on a stand (Fig. 4, left). The study further found that displaying an aggregated visualisation of responses on the urban screen had no improvement on participation rates, unless the screen also showed a live feed of the space around the tablet stand.

Despite the preliminary nature of this field study, the value of the urban screen as enabler for placemaking here is two-fold. First, it contributes to creating a connection between people and the space. This is achieved by allowing passers-by to express themselves through their engagement in an interaction with the screen or tablet. The honeypot effect observed around public displays (Brignull and Rogers 2003) generates additional opportunities for people to connect with each other and to feel more connected with the space. The interaction space provides a stage for people to step in and out of short expressions of public performance (Goffman 1966; Ylipulli et al. 2014). Second, the urban screen empowers people to have some input into long-term placemaking strategies by giving them an opportunity to contribute to urban planning processes. This form of civic engagement ensures that communities can participate in creating and maintaining their distinctive attributes and values (Fredericks and Foth 2013) thereby ensuring that urban planning strategies will lead to more citizen-centric placemaking initiatives.

8 Opportunities and Future Directions

As we have foregrounded in this paper, contemporary manifestations of the built environment such as urban screens, media façades, media architecture, and urban media environments are critical sites for research. This affords an opportunity to consider the notion of the city itself as a living laboratory concerned with developing new visions of what sustainable, urban-centric futures will look like (McArthur et al. 2013). Inevitably, discussions and research into urban screens and media architectures involve the deployment of digital technologies. Fortunately, however, we appear to be moving away from the early emphasis placed on hardware that tended to be couched in terms of the purely technical at the expense of the human experience of living in cities (Yang et al. 2013). More recent understandings of what makes a smart city have focused on expanding definitions of 'smartness' to include knowledge cities, digital cities, and eco-cities (Bell and Niemi-Hugaerts 2013). We have to be ever wary of overstating and idealising what the technologies that drive media informatics can actually deliver and what they might allow citizens and stakeholders to actually achieve.

The potential for innovation in new services, business and models of governance and development are an elusive yet critical focus in the global race to create the 'smartest' cities (Yang et al. 2013). Bernstein et al. (2012) suggest that networks of humans and the computers that drive urban informatics provide a capacity for accessing collective or civic intelligence (Schuler 2009, 2010) in transformative ways useful to education, industry, government, the arts, and the common good. This implies that there are opportunities to reflect and consider the potential of expanded forms of urban media to move beyond the entrancing imagery presented at art and light festivals and on architecture worldwide, and instead trying to engage media savvy citizens in constructive ways (Yang et al. 2013). Simply put, "Smart cities are cities of smart people. We must make sure we are open to finding ways to put this connected intelligence to best use…" (Bell and Niemi-Hugaerts 2013).

Seen in this way, how 'smart' our cities are, is not critically dependent on embedding hardware in concert with new software platforms into the fabric of the built environment. That said, the digital tools and strategies discussed in this paper present cities with an unprecedented array of tools and resources. If deployed in modalities cognisant of local socio-cultural contexts and in ways that seek to be inclusive of the input of citizens we may begin to see smartness reflected in rising levels of social wellbeing and connectedness. These are ultimately the crucial human components in placemaking. Such a holistic and responsive approach to embedding urban screens and media architecture is redolent of what we might describe as a more ecological framework for placemaking. We mean this in the sense that urban media environments will be better positioned to make a viable contribution to the healthy ecology of local urban precincts by enabling us more holistic ways to view, participate in, and understand our complex urban environments. More interdisciplinary research and collaboration in this field will help cities and their stakeholders innovate in the context of creative urban development, community wellbeing, and service provision.

With all the increased attention from researchers, governments, urban planners, architects globally, when we look at the deployment of digital media in urban environments, ideally we should be evolving towards a capacity to envision and build precincts that use media in ways that translate into constructive community and cultural engagement benefiting and empowering beyond the banality of advertising (McQuire 2011) and in so doing enriching the environments, artefacts and public experience of the urban environment (Yang et al. 2013). As noted by McQuire (2011), despite being deployed historically, primarily as a means of advertising or playing television, more recent conceptions of media architecture suggest the potential for developing "…innovative tools for exploring new modes of social interaction and cultural exchange."

Approaches to research and development leading to deeper integration of urban screens and media façades and the creation of viable, vibrant urban media environments that could be considered include:

- Meshing the thinking of artists, designers and other disciplines in contemporary problems associated with precinct development and urban informatics;
- Stimulation of strategies promoting collaboration in 'real' and digital public spaces;
- Integration of online and social technologies as armatures for conceptualisation, communication, collaborative interaction, documentation of ideation, design propositions and processes, and display and archiving;
- Deployment of media artefacts, the body, community, digital networks, public space and architecture as sites for transformations taking into account issues of social construction, politics and sustainability.

9 Conclusions

While many of the smart city initiatives in the past decade have focused on using digital technologies to more efficiently manage cities (Ylipulli et al. 2014), the increasing availability and proliferation of digital screens offer opportunities to add long-term value to urban life by connecting citizens with the city.

Strategies for placemaking that we identified from the analysis of urban screens, media façades, and previous studies, include (a) timing and (b) designing digital screens as participatory platforms. For example, the Midnight Moment event at New York Times Square uses timing to create a unique experience in an urban media environment that is normally dictated by advertising (Haeusler et al. 2012; p. 144). By limiting the daily event to a particular time and timespan, it becomes a special event that connects people within the space as they share this special moment. In a similar way, the popularity of media architecture in light festivals is linked to their temporary nature, making it a destination for people to visit during a particular period during the year, and that way changing the experience of the urban environment.

We discussed two approaches to designing digital screens as participatory platforms in this chapter. First, in situ activism involving digital screens enables people to alter the experience of their built environment (Hespanhol et al. 2014), and second, by allowing people to voice their opinions or contribute to urban planning decisions, using digital screens as communication platform (Schroeter et al. 2012; Valkanova et al. 2014; Hespanhol et al. 2015).

By using such strategies and designing digitally 'augmented' spaces as a whole, which requires the involvement of cross-discipline knowledge, expertise and a shared strategic vision, digital screens can positively reinforce place (Aurigi 2013). This suggests that there is a role for digital screens in urban life that goes not only beyond advertising, but also beyond their use as information displays, which is still a common application in ubiquitous computing research. As research on smart cities and media architecture matures, there is an opportunity to explore how principles from both disciplines can guide the design of digital screens as interfaces that connect citizens with the spaces, in which they live, work, and play.

Acknowledgements The research presented in this chapter was funded through the Henry Halloran Trust at the University of Sydney.

References

Ackad, C., Kay, J., & Tomitsch, M. (2014). *Towards learnable gestures for exploring hierarchical information spaces at a large public display*. Presented at the CHI'14 Workshop on Gesture based Interaction Design: Communication and Cognition, Toronto.

Ackad, C., Wasinger, R., Gluga, R., Kay, J., & Tomitsch, M. (2013). Measuring interactivity at an interactive public information display. In *Proceedings of the Australian Computer-Human Interaction Conference* (OZCHI'13) (pp. 329–332). New York: ACM.

Akpan, I., Marshall, P., Bird, J., & Harrison, D. (2013). Exploring the effects of space and place on engagement with an interactive installation. In *Proceedings of the SIGCHI Conference on Human Factors in Computing Systems* (CHI'13). New York: ACM.

Aurigi, A. (2013). Reflections towards an agenda for urban-designing the digital city. *Urban Design International, 18*(2), 131–144.

Behrens, M., Valkanova, N., Fatah gen Schieck, A., & Brumby, D. P. (2014). Smart citizen sentiment dashboard: A case study into media architectural interfaces. In *Proceedings of the International Symposium on Pervasive Displays* (PerDis'14). New York: ACM.

Bell, H., & Niemi-Hugaerts, H. (2013). Open cities. In K. Braybrooke, J. Nissilä, & T. Vuorikivi (Eds.), *#Reacto 3 The Open Book*. London: The Finnish Institute in London.
Bernstein, A., Klein, M., & Malone, T. W. (2012). Programming the global brain. *Communications of the ACM, 55*(5), 41–43.
Boring, S., Gehring, S., Wiethoff, A., Blöckner, A. M., Schöning, J., & Butz, A. (2011). Multi-user interaction on media facades through live video on mobile devices. In *Proceedings of the SIGCHI Conference on Human Factors in Computing Systems* (CHI'11) (pp. 2721–2724). New York: ACM.
Brignull, H., & Rogers, Y. (2003). Enticing people to interact with large public displays in public spaces. In *Proceedings of the IFIP TC13 Conference on Human-Computer Interaction* (INTERACT'03) (Vol. 3, pp. 17–24). Berlin: Springer.
Bruns, A. (2008). *Blogs, wikipedia, second life, and beyond: From production to produsage*. New York: Peter Lang.
Brynskov, M., Dalsgaard, P., & Halskov, K. (2013). *Understanding media architecture (better): One space, three cases*. Presented at the CHI'13 Workshop on Interactive City Lighting, Paris.
Champion, E., & Dave, B. (2002). Where is this place. In *Proceedings of the International Conference of the Association for Computer-Aided Design in Architecture* (ACADIA'02) (pp. 87–97).
Dourish, P. (2006). Re-space-ing place: "Place" and "space" ten years on. In *Proceedings of the 20th anniversary conference on Computer Supported Cooperative Work* (CSCW'06) (pp. 299–308). New York: ACM.
Foth, M. (2009). Handbook of Research on Urban Informatics: The Practice and Promise of the Real-Time City. IGI Global.
Foth, M., Fischer, F., & Satchell, C. (2013a). From movie screens to moving screens: Mapping qualities of new urban interactions. In J. Geiger, O. Khan & M. Shepard (Eds.), *MediaCity: International Conference, Workshops and Exhibition Proceedings* (pp. 194–204). University at Buffalo, The State University of New York.
Foth, M., Parra Agudelo, L., & Palleis, R. (2013b). Digital soapboxes: towards an interaction design agenda for situated civic innovation. In *Proceedings of the 2013 ACM Conference on Pervasive and Ubiquitous Computing Adjunct Publication (UbiComp '13 Adjunct)* (pp. 725–728). New York, NY, USA: ACM. doi:http://dx.doi.org/10.1145/2494091.2495995
Fredericks, J., & Foth, M. (2013). Augmenting public participation: Enhancing planning outcomes through the use of social media and web 2.0. *Australian Planner, 50*(3), 244–256. (Taylor & Francis).
Frenchman, D., & Rojas, F. (2006). Zaragoza's digital mile: Place-making in a new public realm [media and the city]. *Places, 18*(2).
Fritsch, J., & Brynskov, M. (2011). *Between experience, affect, and information: Experimental urban interfaces in the climate change debate*. From Social Butterfly to Engaged Citizen: Urban Informatics, Social Media, Ubiquitous Computing, and Mobile Technology to Support Citizen Engagement. Cambridge: MIT Press.
Fatah gen Schieck, A. (2009). Towards an integrated architectural media space: The urban screen as a socialising platform. In S. McQuire, M. Martin & S. Niederer (Eds.), *Urban screens reader*. Amsterdam: Institute of Network Cultures.
Fatah gen Schieck, A., Boddington, G., & Fink, P. (2009). *Framework for the implementation of urban big screens in the public space*. London: Bartlett School of Graduate Studies, University College London.
Gianluca, S., Milano, M., Saldivar, J., Nasir, T., Zancanaro, M., & Convertino, G. (2013). Agora 2.0: Enhancing civic participation through a public display. In *Proceedings of the International ACM Conference on Communities and Technologies* (C&T'13) (pp. 46–54). New York: ACM.
Goffman, E. (1966). *Behavior in public places*. New York: The Free Press.
Haeusler, M. H. (2009). *Media façades: History, technology, content*. Germany: Avedition GmbH.

Haeusler, M. H., Tomitsch, M., & Tscherteu, G. (2012). New Media Facades: A Global Survey. Avedition

Hespanhol, L., Tomitsch, M., Bown, O., & Young, M. (2014). Using embodied audio-visual interaction to promote social encounters around large media façades. In *Proceedings of the International Conference on Designing Interactive Systems* (DIS'14) (pp. 945–954). New York: ACM.

Hespanhol, L., Tomitsch, M., McArthur, I., Fredericks, J., Schroeter, R., & Foth, M. (2015). Vote as you go: blending interfaces for community engagement into the urban space. In *Proceedings of the 7th International Conference on Communities and Technologies (C&T '15)* (pp. 29–37). New York, NY, USA: ACM. doi:http://dx.doi.org/10.1145/2768545.2768553

Jacucci, G., Morrison, A., Richard, G. T., Kleimola, J., Peltonen, P., Parisi, L., & Laitinen, T. (2010). Worlds of information: Designing for engagement at a public multi-touch display. In *Proceedings of the SIGCHI Conference on Human Factors in Computing Systems* (CHI'10) (pp. 2267–2276). New York: ACM.

Koolhaus, R. (2014). Fundamentals. 14th International Architecture Exhibition, La Biennale Di Venezia, Marsilio

Korsgaard, H. & Brynskov, M. (2014). City bug report: Urban prototyping as participatory process and practice. In *Proceedings of the International Media Architecture Biennale* (MAB'14). New York: ACM Press.

Kronhagel, C. (2010). *Mediatektur*. Berlin: Springer.

Mayer-Schönberger, V. & Cukier, K. (2013). *Big data—A revolution that will transform how we live, work and think.*

McArthur, I., Murphie, A. & Miller, B. (2013). Augmented_studio: Collaboration, interactive media and urban space. In *Proceeding of the Hybrid Cities Conference*. Athens: University of Athens.

McQuire, S. (2011). Networked urban screens and participatory public spaces. Telecommunications Journal of Australia, *61*(4).

McQuire, S., Martin, M., & Niederer, S. (2009). *Urban screens reader*. INC Reader #5, Institute of Network Cultures.

Memarovic, N., Fatah gen Schieck, A., Schnädelbach, H. M., Kostopoulou, E., North, S., & Ye, L. (2015). Capture the moment: In the wild longitudinal case study of situated snapshots captured through an urban screen in a community setting. In *Proceedings of the 18th ACM Conference on Computer Supported Cooperative Work & Social Computing* (CSCW'15) (pp. 242–253). New York: ACM.

Müller, J., Wilmsmann, D., Exeler, J., Buzeck, M., Schmidt, A., Jay, T., & Krüger, A. (2009). Display blindness: The effect of expectations on attention towards digital signage. *Pervasive Computing* (pp. 1–8). Berlin: Springer.

Ojala, T., Kostakos, V., Kukka, H., Heikkinen, T., Linden, T., Jurmu, M., Hosio, S., Kruger, F., & Zanni, D. (2012). Multipurpose interactive public displays in the wild: Three years later. *Computer* (5), 42–49 (IEEE).

Peltonen, P., Kurvinen, E., Salovaara, A., Jacucci, G., Ilmonen, T., Evans, J., Oulasvirta, A., & Saarikko, P. (2008). It's mine, don't touch!: Interactions at a large multi-touch display in a city centre. In *Proceedings of the SIGCHI Conference on Human Factors in Computing Systems* (CHI'08) (pp. 1285–1294). New York: ACM.

Sarkissian, W., Hofer, N., Shore, Y., Vajda, S., & Wilkinson, C. (2009). *Kitchen table sustainability: Practical recipes for community engagement with sustainability*. London: Earthscan.

Schiffer, M. B., & Miller, A. R. (1999). *The material life of human beings: Artifacts, behaviour and communication*. London and New York: Routledge.

Schroeter, R. (2012). Engaging new digital locals with interactive urban screens to collaboratively improve the city. In *Proceedings of the International Conference on Computer Supported Cooperative Work* (CSCW'12). New York: ACM.

Schroeter, R., Foth, M. & Satchell, C. (2012). People, content, location: sweet spotting urban screens for situated engagement. In *Proceedings of the Designing Interactive Systems Conference (DIS '12)* (pp. 146–155). New York, NY, USA: ACM. doi:http://dx.doi.org/10.1145/2317956.2317980

Schuler, D. (2009). Communities, technology, and civic intelligence. In *Proceedings of the fourth International Conference on Communities and Technologies* (C&T'09) (pp. 61–70). New York: ACM.

Schuler, D. (2010). Community networks and the evolution of civic intelligence. *AI & SOCIETY, 25*(3), 291–307.

Silver, N. (1994). *The making of Beaubourg—A building biography of the Centre Pompidou*. Cambridge: MIT Press.

Taylor, N., Marshall, J., & Blum-Ross, A. (2012). Viewpoint: Empowering communities with situated voting devices. In *Proceedings of the SIGCHI Conference on Human Factors in Computing Systems* (CHI'12) (pp. 1361–1370). New York: ACM.

Tomitsch, M. (2014). Towards the real-time city: An investigation of public displays for behaviour change and sustainable living. In *Proceedings of the 7th Making Cities Liveable Conference, PANDORA Archive*. National Library of Australia.

Tomitsch, M., Ackad, C., Dawson, O., Hespanhol, L., & Kay, J. (2014). Who cares about the content? An analysis of playful behaviour at a public display. In *Proceedings of the International Symposium on Pervasive Displays* (PerDis'14). New York: ACM.

Tscherteu, G. & Tomitsch, M. (2011). *Designing urban media environments as cultural spaces*. Presented at the CHI'11 Workshop on Large Urban Displays in Public Life, May 2011. Vancouver.

Valkanova, N., Jorda, S., Tomitsch, M., & Vande Moere, A. (2013). Reveal-it!: The impact of a social visualization projection on public awareness and discourse. In *Proceedings of the International Conference on Human Factors in Computing* (CHI'13) (pp. 3461–3470). New York: ACM Press.

Valkanova, N., Walter, R., Vande Moere, A., & Müller, J. (2014). MyPosition: Sparking civic discourse by a public interactive poll visualization. In *Proceedings of the International Conference on Computer Supported Cooperative Work* (CSCW'14). New York: ACM.

Vande Moere, A. & Wouters, N. (2012). The role of context in media architecture. In *Proceedings of the International Symposium on Pervasive Displays* (PerDis'12) (p. 12). New York: ACM.

Vlachokyriakos, V., Comber, R., Ladha, K., Taylor, N., Dunphy, P., McCorry, P., & Olivier, P. (2014). PosterVote: Expanding the action repertoire for local political activism. In *Proceedings of DIS'14*. New York: ACM.

Weiser, M. (1991). The computer for the 21st century. *Scientific American, 265*(3), 94–104.

Yang, L., McArthur, I., & Miller, B. (2013). Expanded urban media: From discretized social collages to corrugated social brain. In *Proceedings of the International Symposium for Electronic Arts* (ISEA'13). Sydney: The University of Sydney.

Ylipulli, J., Suopajärvi, T., Ojala, T., Kostakos, V., & Kukka, H. (2014). Municipal WiFi and interactive displays: Appropriation of new technologies in public urban spaces. *Technological Forecasting and Social Change, 89*, 145–160.

Designing Media Architectural Interfaces for Interactions in Urban Spaces

Moritz Behrens, Ava Fatah gen. Schieck and Duncan P. Brumby

Abstract How should one go about designing for interactions with large programmable electronic displays? Part of the challenge is that there are currently only a handful of large interactive surfaces in existence, and so there is much to learn from each attempt to deploy interactive systems. Hence, the work outlined in this chapter contributes to Human–Computer Interaction (HCI) research as well as architectural research by juxtaposing existing interaction frameworks. These frameworks are concerned with the awareness of spaces mediated through information and communications technology (ICT), participants and their actions within these spaces as well as the physical properties of these spaces, which frame these interactions, and are surrounded by the physical built environment. We introduce the notion of media architectural interfaces (MAI), which is then supported through the description of the design, deployment and evaluation of two design studies, namely VEIV London and SCSD Sao Paulo. Finally, we discuss the multilayered interaction frameworks with regard to the conducted design studies and summarize the relevant communalities of these design studies in a taxonomy. The aim of this categorization is to provide design implications for future MAI projects. Ultimately, this may support the design and development of novel and sustainable interactive systems in the domains of urban screens, media facades and media architecture.

M. Behrens (✉) · A. Fatah gen. Schieck
The Bartlett School of Architecture, UCL, London, UK
e-mail: moritz.behrens@ucl.ac.uk

A. Fatah gen. Schieck
e-mail: ava.fatah@ucl.ac.uk

D.P. Brumby
Interaction Centre, UCL, London, UK
e-mail: brumby@cs.ucl.ac.uk

M. Foth et al. (eds.), *Citizen's Right to the Digital City*,
DOI 10.1007/978-981-287-919-6_4

Keywords Public displays · Urban screens · Media facades · Media architecture · Tangible interfaces · Media architectural interfaces

1 Introduction

Cities and their inhabitants are currently familiarizing themselves with the socio-technological developments of the emerging digital age (Hemment and Townsend 2013). Simultaneously, computer technologies increasingly affect architecture in the way that buildings turn into computer interfaces (Mitchell 1996). Today, information and communication technologies (ICTs) may be considered to be of structural, cultural and formal nature (Saggio 2013). Human behaviour is increasingly structured amongst others through information displays and ubiquitous mobile devices characterize our everyday culture. At the same time, architecture is adjusting to the requirements of novel ICT such as large digital displays attached to building facades. For instance, the flashy New York Times Square is an obvious example of where digital media technology has been materialized in architectural form. In essence, the urban environment today can be considered as a system that integrates the human, architecture and ubiquitous computing technologies (Fatah gen Schieck 2006). Within this system are large programmable electronic displays such as urban screens, media facades or media architecture. These novel 'digital' surfaces are gradually turning buildings into responsive media facades, which by using display technologies as an architectural material, radically alter the formal and informational character of the building (Fatah gen Schieck 2006; Ebsen 2013).. Whereas the notion of architectural facades has always been of informational value—although static—in the way that a facade was considered a social interface connecting the bourgeoisie (private space) with the surrounding city (public space) (Neumeyer 2002), the advent of ICTs has allowed for new types of facades to emerge: information is literally not set in stone anymore (Haeusler 2009). Digital information also allows for the anytime exchange of information, such as the flickering advertisement billboards in an inner city, whilst at the same time, the nature of anytime and anywhere information access is changing the relationship between the public and private space. This has led to the development of novel forms of interactions between humans and humans, humans and computers, and humans and architectures. And although Manovich clearly states that the nature of human–computer interactions (HCIs) is interactive per se (Manovich 2001, p. 71), architectural surfaces are by far not.

Having this in mind, one may ask how to go about designing for interactions with large programmable electronic displays? Part of the challenge is that there are currently only a handful of large interactive surfaces in existence, and so there is much to learn from each attempt to deploy interactive systems. Hence, the work outlined in this chapter contributes to HCI as well as architectural research by juxtaposing existing interaction frameworks that are concerned with the awareness of spaces mediated through ICT, participants and their actions within these spaces as well as the properties of these spaces, which frame these interactions, and are surrounded by the physical built environment. We introduce the notion of media architectural interfaces (MAI), which is then supported through the description of the design, deployment and evaluation of two design studies: VEIV London and SCSD Sao Paulo. Finally, we discuss the multilayered interaction frameworks with regard to the conducted design studies and summarize the relevant communalities of these design studies in a taxonomy. The aim of this categorization is to provide design implications for future MAI projects. Ultimately, this may support the design and development of novel and sustainable interactive systems in the domains of urban screens, media facades, and media architecture.

In the next section, we provide an overview of relevant projects and research in the fields of media architecture and human–computer interaction and focus on frameworks that describe social interactions mediated through interactive systems, with a particular focus on urban screens and media facades.

2 Background

2.1 Architecture and HCI

Architectural research, such as Space Syntax research, has dealt with the relationship between architecture and human behaviour. The city is considered as an arrangement of architectural layouts that are defined through their relationships between physical space and social life reflected in movement patterns and activities of its inhabitants (Hillier and Hanson 1984). Space Syntax aims to analyse the spatial morphology of cities through researching the 'relation of space to society, [which] is mediated by spatial configuration. Spatial configuration proposes a theory in which we find pattern effects from space to people and from people to space' (Hillier 1998, p. 301/2). A methodological toolset provided by Space Syntax facilitates the systematic study through spatial analysis and empirical observations of human behaviour such as pedestrian movement or social encounter in the urban realm (Al-Sayed et al. 2013). In this context, social encounters can be seen as (un-)planned gatherings amongst strangers or people who know each other. Architectural research has defined 'shared encounters' as mostly context aware, and hence, the type of encounter stage (e.g. bus stop or museum) and its information context impacts the kind of shared encounters. Encounter stages can be defined as public spaces 'on which people negotiate boundaries of a social and cultural nature' (Fatah gen Schieck et al. 2009, p. 179). For example, at bus stops, social chance encounter happen when people ask for directions or start conversations about the delayed schedule. One objective of the research presented in this paper is to focus on the particular spatial properties, which physical encounter stages require in order to support shared encounters.

Since the advent of ubiquitous computing (Weiser 1991), and its application in urban space in the form of urban computing (Kindberg et al. 2007), the built environment incorporates architecture and ubiquitous computing technologies. In our work, we focus in particular on large electronic displays in public spaces (i.e. public displays), media facades and media architecture and their potential for mediating social interactions. Consequently, architecture and HCI are concerned with the design, deployment and evaluation of digital media technologies and their effect on social interactions in urban spaces. Seen from an architectural perspective, urban screens merged into media facades and became part of the architectural repertoire. Accordingly, media facades are visually animated architectural surfaces, such as dynamic light facades (Virilio 1991; Fatah gen. Schieck 2006; McQuire et al. 2009).

HCI research, on the other hand, is concerned with interactions in general and increasingly with social interactions, which are mediated by computer technologies. More recently, a branch of HCI began to study social interactions and human–computer interactions in the real world (Rogers 2011). Inevitably, this led to a discussion about space within this field. Since then HCI research has conducted extensive research in understanding technology-mediated human behaviour and social interaction in public space (Fischer and. Hornecker 2012; Akpan et al. 2013).

2.2 Media Architecture and the Role of Context

McQuire (2006) argues that TV screens have been transformed from small-scale interior devices to large architectural surfaces that no longer broadcast to private inside spaces but to public outside spaces. Architectural surfaces turned into public media interfaces, transmitting mostly content curated by corporate organisations, rather than interactions generated by the public in situ. The monochromic 'Spectacolour Board' at the New York Times building, set up in 1976, is considered to be the first large electronic display in urban space, which broadcasts dynamic content (McQuire et al. 2009). At the time, this was a technological advancement of the traditional billboard, serving to broadcast commercial content (Huhtamo 2009). From then on, the prevalence of urban screens was unstoppable in particular technological developments and price decline accelerated this trend in recent years. Whole building facades subsequently became digital walls such as the facades surrounding the New York Times Square that display dynamic content.

Urban screens are either stand alone (e.g. large BBC screens initially initiated by the BBC, in Liverpool, UK), attached onto existing building facades (e.g. Piccadilly Circus, London), or the digital media technology is already weaved into the building's surface (e.g. the Galleria department store in Seoul, Korea). Visually animated surfaces, such as dynamic light facades, are equipped with numerous light-emitting diodes (LED). They turn into large programmable pixel matrices displaying animated visual patterns. From a technical perspective, there are other types of artificial light-based media facades as well, such as projections onto facades, back projections through glazed facades, or three-dimensional media facades (i.e. voxel facades) (Haeusler 2009). Some see digital media facades as simple ornaments that create an ambient atmosphere (Caspary 2009). Others consider the potential of digital media facades for communicating content, for instance, advertising (e.g. Times Square, New York or Piccadilly Circus, London), news content (e.g. the network of Big Screens in the UK, which was initially run by the BBC),[1] media art (e.g. Lozano Hemmer's work),[2] social visualization (e.g. BlinkenLights)[3] or for community purposes on a neighbourhood level (e.g. Screens in the Wild).[4]

Extensive research has been carried out to explore the challenges of deploying MAI in public space. Initially, the technical challenges of deploying display technology in public space have been summarized by Streitz et al. (2003). As the design and implementation of digital media facades in the built environment progresses, the purpose of such facades and the contextual characteristics of 'media architecture' are addressed. Parameters that impact the integration of media facades into the existing social fabric from a sociodemographic (environment), technical (content) and architectural (carrier) perspective have been addressed by Vande Moere and Wouters (2012). On the urban scale, the role of space, social proximity and full body performative interactions in shared spaces have been addressed by Fatah gen Schieck et al. (2008), O'Hara et al. (2008) and Peltonen et al. (2008).

2.3 Interfacing with Media Architecture

Due to technological advancement, large public displays became ever more incorporated into the built environment and because of price decline its application for social and artistic purposes became popular in urban space. This led to novel technology-mediated social interactions, such as people engaging with media facades through tangible devices.

Tangible user interfaces (TUIs) 'give physical form to digital information, employing physical artefacts both as representations and controls for computational media' (Ullmer and Ishii 2000, p. 916). Tangible Interactions evolved from research in TUI and rely on embodied interaction, tangible manipulation, physical representation of data and embeddedness in real space and give computational resources and data material form (Hornecker and Buur 2006). The intention is to embed computing into everyday life and support intuitive use.

[1]Big Screens were a collaboration project between the BBC and UK local authorities across 20 different communities. Currently, the Big Screens are run by City Councils.

[2]http://www.lozano-hemmer.com/ (accessed: 30/06/2015).

[3]During the installation, participants were able to create low-resolution light patterns on the façade of an old building through sending SMS via their mobile phones http://blinkenlights.net/ (accessed: 30/06/2015).

[4]http://www.screensinthewild.org/ (accessed: 30/06/2015).

One of the first social media art projects that provided an interface allowing passers-by to create and share content on a media façade was the BlinkenLights[5] project in 2001. Participants on a street in Berlin were able to share messages typed into a mobile phone with the public through posting them on a low-resolution media facade (each pixel was an illuminated window in an office building). Over the past few years there have been numerous media art and research projects that have developed user interfaces and applications to transmit actions, in situ and in real-time, on to a media façade that is connected to an interface. For example, SMSlingshot (Fischer and Hornecker 2012), Sonic Skate Plaza[6] Binoculars,[7] or SCSD Sao Paulo (Behrens and Valkanova 2014).[8]

More recently, multiuser interactions with media facades through mobile devices revealed challenges when deploying interactive artefacts in urban space that enable passers-by to engage with media facades (Boring et al. 2011). Wiethoff and Gehring (2012) introduced a design toolkit to prototype when designing interactions with media facades before the actual deployment. Since then, novel interfaces have been designed and deployed in the urban environment that let people interact with media facades (Hoggenmüller and Wiethoff 2014). We aim to contribute to this existing research by exploring the relation between TUIs and large programmable displays from an HCI as well as architectural point of view.

3 Designing for Interaction: Existing Interaction Frameworks

Weiser (1991) and Dourish (2004) predicted a shift in personal computing and interaction design. Interactions have moved from home and office desktops onto mobile phones, tablets or public displays into urban spaces. Since then, extensive research in HCI has explored social interactions mediated by public displays in public spaces and brought several interaction frameworks forward.

Dalsgaard and Halskov (2010) have outlined the cases and challenges when designing urban media facades and suggest considering eight challenges when designing for media facades. The first two challenges focus on the need for new interfaces that are required in urban settings as well as the integration of these into the existing built environment. Based on this, a framework for designing complex media facades has been developed by Halskov and Ebsen (2013), which includes a description of what the difference is between media facades and conventional displays. Scale, shape of display, pixel shape and their configurations as well as

[5]http://blinkenlights.net/blinkenlights (accessed: 30/06/2015).

[6]http://www.connectingcities.net/project/sonic-skate-plaza (accessed: 30/06/2015).

[7]http://www.connectingcities.net/project/binoculars-to%E2%80%A6-binoculars-from%E2%80%A6 (accessed: 30/06/2015).

[8]http://www.connectingcities.net/project/smart-citizen-sentiment-dashboard (accessed: 30/06/2015).

the quality of light were defined as parameters that differ when comparing media facades with conventional displays. Most obvious is the fact that media facades are not standardized in these properties, the scale is a way bigger than with conventional displays, and compared to conventional displays media facades can have a three-dimensional surface.

In the following, we focus on four existing frameworks that we consider as important milestones towards an integrated media architectural design. Each framework describes the relationship between humans and their action in the presence of programmable electronic displays and in relation to the surrounding space. These frameworks describe the following: (1) awareness space, (2) actor space, (3) action space, and (4) physical space (Fig. 1).

Awareness Space

Research on awareness of public displays in relation to social interactions was first described in HCI research, when noticing a novel social phenomenon around public displays (Brignull and Rogers 2003). Three different types of 'activity spaces' were described: (1) Peripheral awareness activities: activities that take place in the wider space around the display, where people socialize but are not necessarily aware of the presence of the public display. (2) Focal awareness activities: in this space, people are aware of the presence of the display. They are looking at the display, discuss activities that take place around the display or learn how to engage with the content. (3) Direct interaction activities: this is the space where individuals or groups actively engage with the display. The research findings suggest that people found it difficult to transit from one 'activity space' to another.

Later, Vogel and Balakrishnan (2004) published a spatial framework, which described how users fluidly move from implicit interactions in the wider surrounding towards explicit interactions when approaching the direct interaction space around a public display.

Actor Space

The actor space describes the different roles people take on when being in the vicinity of interactive installations. By now, computers have moved away from the desktop and novel interfaces appear, which spread into new spatial settings. People are changing their role; in particular, in public settings people traverse various awareness spaces, which afford specific roles. Consequently, a better

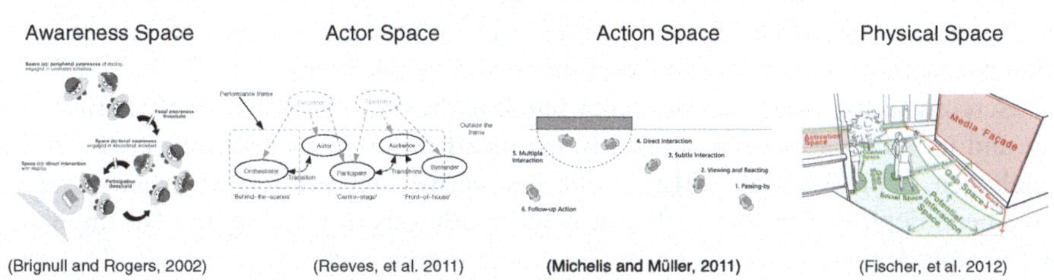

Fig. 1 Four existing sociospatial interaction frameworks

understanding of what kind of roles these are, when and where people take them on, is crucial for the design and deployment of interactive systems in urban spaces. Reeves (2011) identified that the conventional user is actually changing her role when passing-by, on looking or turning into a performer in the vicinity of an interactive installation. Within this framework, the performing user plays the central role when engaging with an interface. Her acting entices other people from the audience into the experience; some of them may become performers. More recently, Behrens et al. (2013), and Fatah gen Schieck et al. (2013), have explored in detail how these roles are framed through the situated layout of urban screens.

Action Space

In the last section, we clarify the diverse roles individuals take on in connection with interactive installations; here, we describe the various phases of interactions people traverse. The Audience Funnel by Michelis and Müller (2011) depicted a framework that establishes a terminology for each transition. These phases were identified as (1) passing by, (2) viewing and reacting, (3) subtle interaction, (4) direct interaction, (5) multiple interactions and (6) follow-up action. Michelis and Müller found that people proceed from one phase to the next in order to understand the interaction. Boundaries are described as a series of thresholds that need to be crossed before one can interact with a public display.

Physical Space

The spatial localization of interactions is largely neglected in the work described above. In contrast, Fischer and Hornecker (2012) have outlined an interaction framework that focuses on the spatial properties of interactions. When looking into the various encounter stages, 'urbanHCI' specifies different 'interaction spaces' on which people perform different activities. This includes the (1) display spaces, (2) interaction spaces, (3) potential interaction spaces, (4) gap spaces, (5) social interaction space, (6) comfort space and (7) activation space on which participants behave differently towards an interactive media facade. Fischer and Hornecker's framework was explored through a media art project called 'SMSlingshot'. This project developed an interface that looks like a wooden slingshot, but its integrated digital technology allows the user to shoot short text messages (SMS) together with virtual paintballs on a media facade. Within these immediate spaces around an electronic display (i.e. interaction spaces), passers-by stop, watch and start playing with the media architectural interface and eventually change the look of the media façade individually; others observe in groups from a distance, discuss or engage as well; simultaneously, other pedestrians do not sense the presence of such interactions and the existence of the media facade at all.

'Screens in the Wild', on the other hand, addressed the question of spatial layout and its relation to technologically mediated interactions (Behrens et al. 2013; Fatah gen Schieck et al. 2013). Four interactive and networked urban screens have been deployed in four different neighbourhoods in London and Nottingham. Within this approach, the following interaction zones were recognized: (1) direct interaction space surrounding the display (direct); (2) the surrounding public space (wide); and (3) across spatial boundaries, i.e. the remotely connected space through the networked displays (networked). Sociospatial configurations mediated

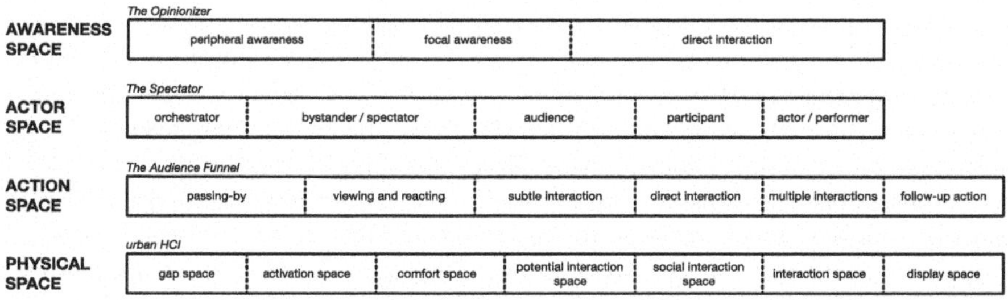

Fig. 2 Comparison of four spatial frameworks for describing interaction mediated through public displays

by urban screens were explored, and site-specific interactions were observed and compared to more generic types of interactions. Indications were found that the properties of the spatial layout play a significant role and, to a certain extent, frame the type of interactions mediated through public displays (Fig. 2).

In summary, we plotted four existing spatial frameworks that describe interactions between humans and displays from (1) awareness spaces, (2) actor spaces, (3) action spaces and (4) physical spaces. Although three of the presented frameworks dealt with smaller public displays, in comparison with the large urban displays, juxtaposing these frameworks assists designers to understand the multilayered design space when designing interactive systems for large programmable displays. We contribute to this body of research by exploring specifically the relation between TUI and large programmable displays (such as media facades) in a given context. We clarify the notion of MAI and apply it on two design case studies we conducted.

4 Media Architectural Interfaces (MAI)

We introduce the notion of MAI. MAI capture an ecology of tangible (TUI) and non-tangible interfaces. They can be considered as interactive systems in urban space, which potentially entice people to step out of their routine and perceive urban space or act differently within it. In more detail, we consider MAI as the synthesis of situated and shared interfaces. They mediate participants' engagement with large programmable electronic displays such as urban screens, media facades or media architecture. Tangible interfaces are generally located on street level, whereas the connected displays are mostly vertical surfaces attached to buildings or are the buildings themselves such as the case with media facades. Eventually, they may disrupt movement and behavioural patterns in the given spatial setting.

The TUIs frame the interaction modalities (Müller et al. 2010) with the display as well as they set the level of participation as described by Fritsch and Brynskov (2011) and complemented by Caldwell and Foth (2014). Usually, these interfaces call for explicit and shared interactions following the 'urbanHCI' framework (Fischer and Hornecker 2012). Large programmable displays amplify participants' interactions through the tangible interfaces, which depend on technical properties of these displays such as type, shape, material, size and resolution of the display (Halskov and Ebsen 2013).

Both interfaces (tangible and non-tangible) are dependent on the given sociospatial setting (for example pedestrianized places, busy high streets or transport hubs). As a consequence, the distance in between the tangible interface and the display can vary. Further, when changing the properties within one of the three constituent elements (i.e. the tangible situated interface, display or setting), the two other elements are directly affected. This will be discussed in more detail in Sect. 5.

In the next section, we report on a case study, which describes an example of a MAI consisting of two similar tangible user interfaces connected to two very different displays within two different sociospatial settings. The aim is to test and explore the notion of MAI and to eventually guide designers when developing interactive systems for urban spaces.

4.1 Case Study

We present two studies: VEIV London and SCSD Sao Paulo. In describing these case studies, we explore the deployment of each MAI and highlight the implications for designing interactions. The focus is on mediated interactions in the vicinity of the deployed MAI. Both case studies differ in the architectural scale and the nature of the urban space (i.e. pedestrianized courtyard vs. congested high street).

VEIV, London
This study (Behrens 2013) is a follow-up on the early implementation we conducted using a TUI of the digital 'I like' button (Fig. 3), situated in the physical

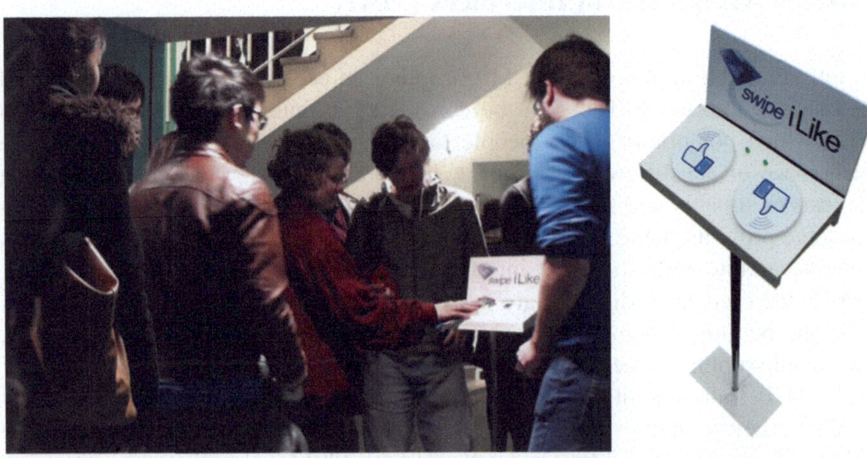

Fig. 3 Initial deployment of the tangible 'I like' button (*right*) on a university campus (*left*)

space within a museum context (Behrens 2011a, b). The findings suggested that visitors are keen to leave their feedback and share it with others (in this case through a simple swipe card interface connecting to Twitter and Facebook). Analysis of the submitted data discovered movement and encounter patterns of visitors interacting with the installed devices and the observations of interactions revealed various forms of human behaviour such as strangers discussing the purpose of the new interfaces. In other words, the space around the tangible device turned into an encounter stage that did not exist at this place before. However, it became clear that a real-time visualization or representation of the gathered data shown to the audience as an immediate feedback of their actions was desired by many participants but missing in this initial implementation.

Building on these initial findings, the purpose of the VEIV London study was to: first, explore whether a feedback device that works in a condensed indoor public space can also be deployed in an outdoor public setting full of urban distractions, and second, add a dynamic display to the device to fulfil the need of providing the user with an immediate visual feedback. Consequently, we ask what kind of social interactions and dynamic relations of human behaviours take place in an urban setting when mediated by this interactive installation.

The study took place during the UCL Virtual Environments, Imaging and Visualisation doctoral centre (VEIV) festival, 2013, where we set up the first prototype of the LED display at the UCL main campus. To explore interactivity, we connected the light installation to the binary feedback device based on Radio Frequency IDentification (RFID) technology, which was used in the indoor public space of the museum (Behrens 2011b). During the festival, participants were able to leave their feedback in a playful way through swiping their travel cards (Oyster) or UCL access cards over the thumbs-up or thumbs-down icons on the card reader. Basically, the simple but effective question was whether visitors like or don't like the VEIV Centre. When swiping across the thumbs-up icon, the light installation turned the VEIV logo into a warm orange, whilst the thumbs-down icon turned the logo into a cool blue (Fig. 4).

Fig. 4 LED installation with participant interacting through a binary RFID feedback device. Comfortable *orange colour* represent the thumbs-up icon on the (*left*) and the *coldish blue colour* (*right*) was linked to the thumbs-down button

The LED display consists of three low-resolution displays following the principle of 16 digit number displays, which allow the projection of numbers and letters. In total, 48 RGB 24V LED light strips are connected to microcontrollers, which are addressable through the DMX protocol. A decoder transfers the DMX signals through USB wire to a laptop. The visualizations are programmed with processing. The installation was running between 8 and 10 pm on a summer day. During the first hour, the daylight reduced the light distribution of the LED light, whereas with the approaching darkness the LEDs were colouring the surrounding buildings in either orange or blue ambient light.

Throughout the event, we took pictures, notes and informally talked to people joining the event. Overall, participants enjoyed using their Oyster Cards or UCL access cards to change the colour of the low-resolution display. The fact that they were actually rating the event was less important than the playfulness of changing the colours.

We observed people sitting on the lawn suddenly walking up to change the colour from the apparently uncomfortable blue to the warm orange. They obviously felt more comfortable with the cosy orange light than with the cold blue light. A kind of campfire atmosphere was created where people were sitting around a source of pleasant light that illuminates the faces of others as well as putting the surrounding facades of the classic campus building into an orange-red shade (Fig. 5 (1–6)).

The lawn in front of the installation was the preferred seating area and the TUI (i.e. 'I like' device) created a stage for social encounters. However, after sunset, people moved away from the immediate space around the display as the LEDs became too bright. Interestingly, people neither sat nor stood in between the tangible interface and the LED display nor behind the LED display(Fig. 5).

The DIY and low-resolution media display worked out very well considering the relatively low effort that was put into the design and making of the low-resolution screen and the visual impact of illuminating a courtyard during a large public event was huge. The fact that the lightweight frames are easily movable makes this project easily adaptable to other locations.

SCSD, Sao Paulo

The Smart Citizen Sentiment Dashboard (SCSD) is an interactive participatory installation that lets citizens engage with, and comment on, urban challenges in their cities. Through a tangible interface connected to a media facade, passers-by and participants on-site can submit their sentiments and simultaneously see the effect of their actions projected onto the facade. The tangible urban interaction device allows for an intuitive and accessible, yet identifiable and public way of expressing ones view. The project aims to create an open, aesthetic dialogue about urban challenges and invites citizen to engage, by playfully allowing them to express their opinion and share and compare their views in the physical built environment.

Our motivation was to design, develop and deploy a situated system that mediates collaborative interactions in public spaces, whilst focusing on accessibility and affordance. In other words, the interface should be understandable and easy

Fig. 5 (*1*) Three low-resolution LED frames connected to a tangible 'I like' button (TUI); participants (standing) gather around the TUITUI and talk to each other; (*2*) visitors cue for beverages and food whilst watching the installation (in the background); visitors (in the foreground) are sitting in groups on the lawn; (*3*) after sunset, the outline of the frames disappeared and the VEIV font became visually dominant; the 24V LEDs are very bright and wash the surrounding in *cool blue light*; people still sitting in groups in front of the display; (*4*) some people enjoy sitting in the direct vicinity of the display seem to feel uncomfortable with the bright LED lights and turned their backs towards the frames; (*5*) throughout the evening the space in front of the display turned into a social encounter stage were people mingled, unexpectedly met friends or got to know new people; (*6*) until late groups were gathering in front of the display and seemed to enjoy the cosy campfire atmosphere offered the *orange* LED light

to use for people. Based on the feedback of the previous pilot study, described in the previous section, we deployed and tested-out the TUI in various occasions and improved the system iteratively. Similar to the VEIV London study, the employed technology makes use of existing RFID technology as known from smart card technology. We build on the widely spread use of these unique ID tags for payless travel purposes or building access, as a large proportion of city dwellers carries a RFID tag in their pocket (i.e. the Bilhette Unico is the transport card in Sao Paulo). Consequently, the use of these cards is a recurring embodied interaction in the technologically augmented city. At the same time, every interaction is uniquely identifiable and therefore traceable. Our aim was to allow people to use their ID tags beyond technical purposes and express their mood and opinion about specific issues in the technology-mediated urban realm. Hence, the Smart Citizen Sentiment Dashboard (SCSD) (Fig. 6, left) enables participants to express their mood about urgent urban challenges in the city of Sao Paulo (Behrens et al. 2014).

Up front, we were running four design ethnographical workshops amongst various social groups in Sao Paulo with the aim to learn about citizens' urban challenges. As a result of the collaboration, five categories were established: (1) environment, (2) mobility, (3) security, (4) public space and (5) housing. By switching a knob on the device, participants are able to choose one of the

Fig. 6 (*Left*) The Smart Citizen Sentiment Dashboard (i.e. TUI) connected to the large media façade (i.e. display) of the FIESP building on Avenida Paulista in Sao Paulo, Brazil (2013); (*Right*) people at a nearby bus stop watching the dynamic media façade whilst waiting for the bus

aforementioned categories. By swiping their RFID token (i.e. Bilhette Unico) across one of the three emotions (happy, indifferent and sad), their mood was transmitted on to the media façade (i.e. display) (Fig. 3).

The mood expressed by the user (i.e. happy, indifferent or sad) is then projected onto a huge LED media façade (i.e. display), which has been retrofitted in the existing honeycomb façade of the pyramidal FIESP building.

The media facade is divided into three parts, which are situated on three different sides of the facade. The biggest and main display faces to the opposite side of the street, whereas the two smaller screens are directed to display to both directions of Avenida Paulista. The threefold low-resolution LED facade is formed of a network of approximately 26,000 LED clusters (pixels) embedded in 3700-m^2 metal structure that covers the pyramidal FIESP building. The grid is approximately 13 × 13 cm. Each pixel consists of a module of four LEDs: 2 × R, 1 × G, 1 × B with a luminous intensity of 4.5 cd/module.

For the display, we chose a visualization technique that combines the 'seriousness' of the topic with the more accessible style of popular info-graphics. The visualization comprises of an abstract sunburst representation, of which each burst corresponds to the sentiment of an individual participant towards the currently selected urban challenge. Each urban challenge is encoded by a different colour and an icon representation. Upon switching the rotary knob (Fig. 6), the sunburst visualization corresponding to the specific urban challenge and coloured accordingly appears on the facade. The sentiment 'value' for each participant (happy, indifferent, sad) is graphically encoded through the length of the corresponding burst: the longest burst represents a positive sentiment towards the urban challenge at hand, whilst the shortest corresponds to a negative statement. Our choice for this circular visualization technique was also motivated by its scalability, which allows for an arbitrary number of people to participate and be visually represented.

We considered this flexibility a desirable feature in the context of urban environments, often characterized by highly variable and open-ended, and unpredictable flux of people and interactions. The integration of dynamic visual cues can make visualizations richer, more vivid and therefore easier to understand. Accordingly, our visualization shows a dynamically animated circle over the sunbursts in order to convey the average participants' sentiment for the given urban challenge. Each new burst from a participant visually appears with a smooth animation and bouncing effect, to highlight the recording of fresh data. A new entry is displayed in a white colour to unambiguously distinct it from the rest of the graphical representation. Shortly after, it is smoothly taken over by the colour of its respective urban challenge. In order to provide citizens with an overview of previously submitted sentiments, and with a more interactive approach to exploring the installation, we integrated a 'heart' button at the bottom of the interface (Fig. 7). When pushing this button, a dynamic visualization of the average feedback for all available urban challenges is represented on the facade. As mentioned above, each urban challenge is represented by its corresponding colour and occupies a different part of the circular shape proportionally to the relative participation rate of the according challenge. We aimed to create a simple, playful, yet meaningful approach to enable citizens and participants alike to make a deeper sense of the installation, and the underlying participation results: people can gain insight about which urban challenge is most attractive to vote for, and what is the average sentiment about it of fellow citizens. This, beyond being an overview, the heart visualization symbolizes the overall 'sentiment' of the city towards its urban challenges.

Fig. 7 Three-fold interactions: (1) *switching*: 5 categories can be selected through a rotary switch; (2) *swiping*: after choosing the category, the electronic ID card needs to be swiped over one of the three mood states (happy, indifferent, sad); (3) *pushing*: finally, a simple push button allows users to view the overall feedback of all collected moods

5 Findings and Discussion

Our findings suggest that the implementation of a tangible interface not only supports engagement indoor, such as in a museum context, but also outdoor during a festive event and a media art festival. Yet, the actual TUI in an outdoor setting was not immediately attracting participants. Instead, our observations suggest that the strong visual presence of the display (i.e. low-resolution display or media facade) appealed peoples' attention first.

In the first study, VEIV London, the interactive system consisted of a simple binary RFID swipe card interface. In the second study, conducted in Sao Paulo, the shared user interface was again based on RFID technology but with additional features such as a rotary knob and a push button. Initially, we decided to use RFID technology instead of simple touch buttons due to the potential to identify the user. We were able to track individual user behaviour such as returning users or misapplication, which simple buttons would not allow. Despite these useful features, we observed that the need for an individual swipe card generated a barrier that was hindering some people to interact. The reasons for this may be manifold: in London, the use of the transport card (i.e. Oyster Card) has a wider distribution amongst all citizens. Public transport is generally considered to be a convenient way of commuting and people feel secure using London buses or undergrounds. Consequently, the usage of the Oyster Card appears to be more embodied than in Sao Paulo, where private transportation is very common, mostly due to personal security. Thus, Londoners more likely understand how to engage with the interface and particularly liked the idea to use a smart card to give feedback about services. However, we observed that the fact that not all people carrying a RFID card in their pockets triggered unexpected social encounters during both studies. Bystanders, curious to get involved, asked actors to borrow their cards, or families sharing one card amongst each other. In addition, bystanders started debating about the sense of having such interfaces in urban space or simply wanted to know how the technology works.

During the SCSD Sao Paulo study, we looked closely at actors directly interacting with the TUI. The most common behaviour actors performed started with looking at the TUI, swiping their card or turning the rotary knob. After expressing their feelings towards one of the categories at the swipe card interface, we saw that participants would then frequently lookup to the media façade to see what impact their acting had on the visualization. The fact that the visualization was slightly delayed (i.e. less than a second) was an advantage for the actor experience as it left time for the body orientation towards the large media façade. Close bystanders were behaving almost synchronic in order to understand the interaction modality (Fig. 8).

At the same time, we made observations in the wider surrounding of the MAI during the SCSD Sao Paulo deployment. We recognized MAI-related interactions around the façade in the close interaction space as well as in the wider ambient space. Although these observations were not rigorously conducted, we did notice

Fig. 8 When participants approach the device, they first chose one of the five categories (mobility, public space, housing, safety, environment) by switching manually the rotary knob (*far left*); they then double check if the façade changed accordingly (*left*); after this, they use their RFID card to express their opinion about the urban challenges by swiping the card (*right*); they check again how the façade reacted to their actions (*far right*)

a few recurring behaviours. In particular, we frequently saw people taking pictures of the media façade with their mobile phones or taking pictures of each other in front of the façade. These informal observations would suggest that people simply liked the visual presence of the media façade's visualization and the heart icon it used. Yet, the initial design aim was to develop an interactive system, which would spark passers'-by engagement with the informational content and eventually may trigger new social encounters and discussions about the matter. In the case of VEIV London, participants were asked to express their opinion about the UCL doctoral centre (i.e. like or don't like), whilst during the SCSD Sao Paulo event contestants could share their sentiments about urgent urban challenges in their city (i.e. environment, mobility, security, public space or housing). Accordingly, this raises the question regarding the informational character of the installation as intended and the actual ambient perception of the visualization by many people in particular outside the direct interaction space in the close vicinity of the TUI. Our findings here suggest that designers need to take the ambient perception of MAI set ups into account when designing interactive systems in urban environments (Fig. 9).

More specifically, we would like to relate our findings to the initially examined interaction frameworks in Sect. 3, which dealt with the (1) awareness space, (2) actor space, (3) action space and (4) physical space in the context of public displays or a media façade. In the following, we will discuss these frameworks in relation to the two design studies as outlined in the last section. We argue for the consideration of these multilayered interaction spaces when designing interactive systems in urban spaces.

Fig. 9 Ambient experience in the wider space around the display: passers-by stop and take pictures of each other (*right*); onlookers waiting at the pedestrian crossing, taking pictures with mobile phones and sharing them (*left*)

Awareness Space

In both studies (VEIV London and SCSD Sao Paulo), the direct interaction activities took place in the immediate vicinity of the TUI, whereas the focal and peripheral awareness spaces differed in their dimensions. This is firstly due to the dissimilar size of the displays. The display used for the VEIV London study was only 2 by 6 m (i.e. 12 m^2) compared to the large media façade for the SCSD Sao Paulo study with 3700 m^2. At the same time, the difference in distance between the user interface and the display was significant: 5 m VEIV London compared to 33 m SCSD Sao Paulo. This seemed to have an impact on people's spatial awareness with regard to dynamic displays in the city.

Actor Space

Compared to the dynamic stream of pedestrians in the SCSD Sao Paulo study, the flow of people in the VEIV London study was rather calm, whilst the event was highly attended. The audience hardly changed during the party, but their actions altered; for instance, spectators turned into actors when walking up to the TUI to change the colour of the display and returned to their friends. Others simply enjoyed the presence of the colourful display or watched other guests interacting with it. In contrast, the dwelling time of the audience during the Sao Paulo event was much lower due to the high pass-through frequency of pedestrians on their way to the underground station.

Action Space

The flow of people's actions during the VEIV London event was different to SCSD Sao Paulo. Other than Michelis and Müller (2011) stated, the phases through which people had to go before they could actually interact with the MAI were not stringent in the VEIV London set up. People did not necessarily cross in the described order from (1) passing by to (2) viewing and reacting to (3) subtle interactions to (4) direct interactions to (5) multiple interactions and to (6) follow-up actions. Instead, we observed groups gathering on the lawn walking up to the TUI to collectively explore the interactive system. In contrast to this, we identified human behaviour as explained by Michelis and Müller (2011) during the SCSD Sao Paulo study.

Physical Space

The spatial layout in the VEIV London case consisted of a pedestrianized rectangular courtyard enclosed by the historical university buildings, compared to the vibrant two-directional avenue in Sao Paulo, which was lined by high-rise office towers. In the VEIV London set up, there was hardly any gap space in between the TUI and the display (i.e. ca. 5 m), compared to the congested main road in Sao Paulo, which divided the tangible interface from the media façade (i.e. ca. 33 m). In addition, the comfort spaces, potential interaction spaces and social interaction spaces varied greatly in both studies. The enclosed courtyard and the positioning of the display within it released more of these spaces than the congested pavement in Sao Paulo.

In summary, it appears that the affordance of each of the described spaces has a strong impact on the presented studies and therefore need to be taken into account when designing for such interactive systems in the built environment. With this in mind, we return to our initial question: How to actually go about designing for an interaction with MAI? To help answering this question, we now introduce a taxonomy for the categorization of MAI based on the triangular relationship between TUI, display and setting. Within Fig. 10, we have plotted the relevant information of both design studies against each other to gain a better understanding of the relation between the individual properties.

In more detail, the properties of each constituent element will be described below:

Interface

The characteristics for the participation level have been described by Fritsch and Brynskov (2011) as (1) static, (2) dynamic, (3) reactive, (4) interactive,

constituent element	display							interface			context	
properties	type of display	material	shape of display	size of display	distance between interface and display	resolution of display	content	type of interface	participation level	interaction modalities	social setting	spatial setting
characteristics							ambient, animated, static, dynamic	situated, embedded, embodied	static, dynamic ,reactive, interactive, participatory, communicative, performative, controllable	presence, body position, body posture, facial expression, gaze, speech, gestures,remote control, Keys, touch		
VEIV London	mobile DIY display	RGB LED stripes on wooden frame	rectangular display	12 m ² 2m x 6m	ca. 5m	low	feedback sharing dynamic visualisation; ambient	situated RFID swipe interface	dynamic, interactive, participatory, communicative, controllable	remote control, touch	university event	pedestrianised court / leisure area
SCSD Sao Paulo	permanent, retrofitted media facade	individual RGB LED modules fitted in honeycomb facade	three display sides	3700 m ²	ca. 33m	medium	sentiment data dynamic data visualization; animated	situated RFID swipe interface; switch and button controls	dynamic, interactive, participatory, communicative, controllable	remote control, keys, touch	media art festival	busy high street / transit space

Fig. 10 Taxonomy towards an integrated media architectural design space following the notion of MAI

(5) participatory and (6) communicative and recently have been extended by Caldwell and Foth (2014), who added the terms (7) performative and (8) controllable. The characteristics of the interaction modalities originate from dimensions summarized by Müller et al. (2010) and use the following interaction modalities: (1) presence, (2) body position, (3) body posture, (4) facial expression, (5) gaze, (6) speech, (7) gestures, (8) remote control, (9) keys and (10) touch. This may also impact the distance between the interface and the display. Being aware of these properties and their characteristics will allow designers clarifying the nature of interaction they want to design for.

Display
These properties are mostly related to the technical details of large programmable displays as described by Halskov and Ebsen (2013). They consist of (1) type, (2) material, (3) shape, (4) size and (5) resolution but also of (6) the type of content. Each of these characteristics can impact the multilayered spatial frameworks described above (i.e. awareness, action, actor, physical).

Context
Understanding the sociospatial settings as described in Sect. 2.1 are core properties when locating a MAI. As explained in this paper, here is a huge difference when designing in the context of a pedestrianized university court (i.e. VEIV London) or a busy high street (i.e. SCSD Sao Paulo).

Although most of the characteristics aligned in our taxonomy describe the properties of the constituent elements (i.e. display, interface, context) from a technical point of view, they influence the multilayered spatial frameworks (i.e. awareness, actor, action, physical) and consequently assist when designing MAI for interactions in urban space. Through understanding the effect of the single properties of each constituent element on the whole interactive system, designers may eventually be able to be fully aware of their design decisions.

6 Conclusion

In this paper, we ask the question 'how to go about designing for an interaction with such a large programmable electronic display' and introduced the notion of MAI founded on existing spatial interaction frameworks established in HCI research in recent years. We further described two design studies, which we designed, deployed and observed in urban space. The first study was carried out at a gathering in a university courtyard (i.e. VEIV London), whilst the second was conducted in a complex urban setting in the city of Sao Paulo (i.e. SCSD Sao Paulo). Further, we described the setting in which the interactive system was arranged and presented our observations concerning mediated social interactions in the surrounding space. Based on our findings, we discussed the relation of our observations to four existing spatial frameworks and suggested a taxonomy, which categorizes the properties and characteristics related to our notion of MAI.

In this work, we laid the focus on social interactions mediated through our MAI. For future development, we aim to explore other MAI, study their properties between human behaviour, the user interface and the display in a given spatial layout and compare them with our taxonomy.

References

Akpan, I., Marshall, P., Bird, J., & Harrison, D. (2013). Exploring the effects of space and place on engagement with an interactive installation. In *Proceedings of the SIGCHI Conference on Human Factors in Computing Systems (CHI '13)*. New York, NY, USA: ACM.

Al-Sayed, K., Turner, A., Hillier, B., Iida, S., & Penn, A. (2013). *Space syntax methodology*. London: Bartlett School of Architecture, UCL.

Behrens, M. (2011a). Swipe 'I like': Location based digital narrative through embedding the 'Like' button in the real world. In *Presented at: 5th International Conference on Communities & Technologies—Digital Cities 7*.

Behrens, M. (2011b). Can we enhance the engagement of online and real world communities with museum content through embedding a location-based 'I like' button? *Thesis for M.Sc. Adaptive Architecture and Computation*. London: Bartlett School of Graduate Studies, University College London.

Behrens, M. (2013). Living light lab: Exploring instant feedback in mediated urban space. In *UbiComp '13 Adjunct*, Zurich, Switzerland, September 08–12, 2013.

Behrens, M., Fatah gen. Schieck, A., Kostopoulou, E., North, S., Motta, W., Ye, L., & Schnädelbach, H. (2013). Exploring the effect of spatial layout on mediated urban interactions. In *Proceedings PerDisp'13: The International Symposium on Pervasive Display, California, USA*.

Behrens, M., Valkanova, N., Fatah, A., & Brumby, D. (2014). Smart citizen sentiment dashboard: A case study into media architectural interfaces. In *Proceedings, PerDis'14: The International Symposium on Pervasive Display, Copenhagen, Denmark*. New York: ACM library.

Boring, S., Gehring, S., Wiethoff, A., Blöckner, A. M., Schöning, J., & Butz, A. (2011). Multi-user interaction on media facades through live video on mobile devices. In *Proceedings of the SIGCHI Conference on Human Factors in Computing Systems, CHI '11* (pp. 2721–2724). New York, NY, USA: ACM. doi:10.1145/1978942.1979342

Brignull, H., & Rogers, Y. (2003). Enticing people to interact with large public displays in public spaces. In *Proceedings of INTERACT* (Vol. 3, pp. 17–24).

Caldwell, G. A., & Foth, M. (2014). DIY media architecture: open and participatory approaches to community engagement. In *Proceedings of the 2nd Media Architecture Biennale Conference: World Cities (MAB '14)* (pp. 1–10). New York, NY, USA: ACM. doi:http://dx.doi.org/10.1145/2682884.2682893

Caspary, U. (2009). Digital media as ornament in contemporary architecture facades: Its historical dimension. In S. McQuire, M. Martin, & S. Niederer (Eds.), *Urban screens reader*.

Dalsgaard, P., & Halskov, K. (2010). Designing urban media façades: Cases and challenges. In *Proceedings of the 28th International Conference on Human Factors in Computing Systems* (pp. 2277–2286).

Dourish, P. (2004). Where the action is: The foundations of embodied interaction (New Ed.). Cambridge: MIT Press.

Ebsen, T. (2013). *Material screen: Intersections of media, art, and architecture*. Denmark: Aarhus Universitet.

Fatah gen Schieck, A. (2006). Towards an integrated architectural media space. Special Issue #4: Urban screens: Discovering the potential of outdoor screens for urban society. *First Monday, 11*(4).

Fatah gen Schieck, A., Al-Sayed, K., Kostopoulou, E., Behrens, M., & Motta, W. (2013). Networked architectural interfaces: Exploring the effect of spatial configuration on urban screen placement. In *Proceedings of the 9th International Space Syntax Symposium, Seoul, South Korea*.

Fatah gen Schieck, A., Briones, C., & Mottram, C. (2008) The urban screen as a socialising platform: exploring the role of place within the urban space. In *MEDIACITY. Situations, practices and encounters*. Germany: Frank & Timme GmbH.

Fatah gen Schieck, A., Kostakos, V., & Penn, A. (2009). Exploring digital encounters in the public Arena. In K. S. Willis, G. Roussos, K. Chorianopoulos, M. Struppek (Eds.), *Shared encounters [Internet]* (pp. 179–195). London: Springer. Available from http://link.springer.com/chapter/10.1007/978-1-84882-727-1_9

Fischer, P. T., Hornecker, E. (2012). Urban HCI: spatial aspects in the design of shared encounters for media facades. In *Proceedings of the SIGCHI Conference on Human Factors in Computing Systems, CHI '12* (pp. 307–316). New York, NY, USA: ACM.

Fritsch, J., & Brynskov, M. (2011). Between experience, affect, and information: Experimental urban interfaces in the climate change debate. In M. Foth, L. Forlano, C. Satchell, & M. Gibbs (Eds.), *From social butterfly to engaged citizen* (pp. 115–134). Cambridge, MA: MITP.

Haeusler, H. (2009). *Media facades: History, technology, Content*. Ludwigsburg: AVEdition.

Halskov, K., & Ebsen, T. (2013). A framework for designing complex media facades. *Design Studies, 34*(5), 663–679. doi:10.1016/j.destud.2013.04.001

Hemment, D., & Townsend, A. (2013). *Smart citizens*. Manchester: FutureEverything Publications.

Hillier, B. (1998). *Space is the machine: A configurational theory of architecture*. Cambridge: Cambridge University Press.

Hillier, B., & Hanson, J. (1984). *The social logic of space, Bd.* Cambridge: Cambridge University Press.

Hoggenmüller, M., & Wiethoff, A. (2014). LightSet: Enabling urban prototyping of interactive media façades. In *Proceedings of the 2014 Conference on Designing Interactive Systems, DIS '14* (pp. 925–934). New York, NY, USA: ACM. doi:10.1145/2598510.2598551

Hornecker, E., & Buur, J. (2006). Getting a grip on tangible interaction: A framework on physical space and social interaction. In *Proceedings of the SIGCHI Conference on Human Factors in Computing Systems* (pp. 437–446). New York: ACM.

Huhtamo, E. (2009) Messages on the wall: An archaeology of public media displays. In S. McQuire, M. Martin, S. Niederer (Eds.), *Urban screens reader*.

Kindberg, T., Chalmers, M., & Paulos, E. (2007). Guest editors' introduction: Urban computing. *Pervasive Computing, IEEE, 6*(3), 18–20.

Manovich, L. (2001). *The language of new media*. Cambridge: MIT press.

McQuire, S. (2006). The politics of public space in the media city. Special Issue #4: Urban Screens: Discovering the potential of outdoor screens for urban society *First Monday* (February 2006), http://dx.doi.org/10.5210/fm.v0i0.1544.

McQuire, S., Martin, M., & Niederer, S. (Eds.). (2009). *Urban screens reader*. Amsterdam: Institute of Network Cultures.

Michelis, D., & Müller, J. (2011). The audience funnel: Observations of gesture based interaction with multiple large displays in a city center. *International Journal of Human-Computer Interaction, 27*(6), 562–579. doi:10.1080/10447318.2011.555299

Mitchell, W. J. (1996). City of bits: Space, place, and the infobahn (new ed.,). Cambridge: MIT Press.

Müller, J., Alt, F., Schmidt, A., & Michelis, D. (2010). Requirements and design space for interactive public displays. In *Proceedings of the International Conference on Multimedia* (pp. 1285–1294). doi:10.1145/1873951.1874203

Neumeyer, F. (Ed.). (2002). *Quellentexte zur Architekturtheorie*. New York: Prestel.

O'Hara, K., Glancy, M., & Robertshaw, S. (2008). Understanding collective play in an urban screen game. In *Proceedings of CSCW 08, San Diego, CA, USA*.

Peltonen, P., Kurvinen, E., Salovaara, A., Jacucci, G., Ilmonen, T., Evans, J., & Saarikko, P. (2008). It's Mine, Don't Touch!: Interactions at a large multi-touch display in a city centre. In *Proceedings of the SIGCHI Conference on Human Factors in Computing Systems* (pp. 1285–1294). New York: ACM.

Reeves, S. (2011). *Designing interfaces in public settings: Understanding the role of the spectator in human-computer interaction*. Berlin: Springer Science & Business Media.

Rogers, Y. (2011). Interaction design gone wild: Striving for wild theory. *Interactions, 18*, 58–62.

Saggio, A. (2013). *The IT revolution in architecture. Thoughts on a Paradigm Shift*. New York: Lulu.com.

Streitz, N., Prante, T., Röcker, C., Van Alphen, D., Magerkurth, C., Stenzel, R., & Plewe, D. A. (2003). Ambient displays and mobile devices for the creation of social architectural spaces. In *Public and situated displays* (pp. 387–409). The Netherlands: Springer.

Ullmer, B. & Ishii, H. (2000). Emerging frameworks for tangible user interfaces. *IBM Systems Journal, 39*(3.4), 915–931. Available at http://ieeexplore.ieee.org/lpdocs/epic03/wrapper.htm?arnumber=5387042

Vande Moere, A. & Wouters, N. (2012). The role of context in media architecture. In *Proceedings PerDisp'12*. New York: ACM.

Virilio, P. (1991). *The lost dimension*. New York, N.Y.: Semiotext(e).

Vogel, D., & Balakrishnan, R. (2004). Interactive public ambient displays: Transitioning from implicit to explicit, public to personal, interaction with multiple users. In *Proceedings of the 17th Annual ACM Symposium on User Interface Software and Technology* (pp. 137–146).

Weiser, M. (1991). The computer for the 21st century. *Scientific American, 265*(3), 94–104.

Wiethoff, A., & Gehring, S. (2012). Designing interaction with media façades: A case study. In *Proceedings of the Designing Interactive Systems Conference* (pp. 308–317). New York: ACM.

Design Patterns for Urban Gardening

Peter Lyle, Marcus Foth and Jaz Hee-jeong Choi

Abstract Urban agriculture (UA) is practised in many forms within urban spaces, ranging from large organised community gardens to individuals with a backyard or balcony. We present the synthesis of findings from three studies exploring different types of UA: city farms, residential gardeners, and a grass roots group that supports local communities. Where the findings of individual studies are used to justify a design approach, there are often difficulties encountered because of different context of the original study. Through our understanding and synthesis of multiple studies, we propose a set of design patterns. The proposed patterns can be utilised concurrently depending on the scale and context of both the physical garden and community. The relationships between the patterns and their concurrent use are discussed, and the resulting links provided the foundation for a pattern language. The eight initial patterns provide a foundation on which we would encourage other researchers to contribute, in order to develop a pattern language to holistically consider UA beyond the scope of our experiences in Brisbane and to enrich the patterns with a variety of gardening practices.

Keywords Urban agriculture · Interaction design · Gardening · Food · Design patterns · Pattern language · Urban informatics

P. Lyle (✉) · M. Foth · J.H. Choi
Urban Informatics Research Lab, Queensland University of Technology,
Brisbane, QLD, Australia
e-mail: p.lyle@qut.edu.au

M. Foth
e-mail: m.foth@qut.edu.au

J.H. Choi
e-mail: h.choi@qut.edu.au

M. Foth et al. (eds.), *Citizen's Right to the Digital City*,
DOI 10.1007/978-981-287-919-6_5

1 Introduction

Recently, there have been a number of studies exploring the intersection of urban informatics and urban agriculture (UA) communities. UA offers a variety of benefits to practitioners in the form of mental and physical health (Bellows et al. 2003), improved food security to the community (Kortright and Wakefield 2011), and a reduction in energy requirements to process, store, and transport produce (Bellows et al. 2003; Schnell 2013). The impact of UA on society is becoming more important as the level of urbanisation increases; for example, in Australia where we have conducted studies with local urban gardeners, the urban population constitutes 89 % (United Nations 2014). Urban informatics refers to study, design, and practice of urban experiences, using technology that connects people to each other, both in the physical and digital spaces (Foth et al. 2011). With the increasing ubiquitous nature of interactive technology, urban informatics research has been exploring the opportunities and challenges that this presents.

Pattern languages and design patterns (the constituent parts) have been of interest to a number of areas within human–computer interaction (HCI), although originating in the field of architecture. Based on three studies, we have conducted with different types of UA in Brisbane, Australia, between 2010 and 2014, and supported by existing studies, we propose an initial set of design patterns for urban gardening as practical guidelines for the design of interactive technologies for UA communities. We intend these design patterns to provide for a foundation for a future pattern language, and moving forward, we encourage other researchers to contribute new patterns, and test and improve the current patterns.

We give an overview of the use and role of pattern languages in architecture and civic engagement, as well as different uses within ICT and HCI research. We then provide an overview of the three studies from which we have analysed for the purpose of the developing our patterns. We then discuss the theoretical implications of pattern languages as an approach to generalising findings, before outlining our method of analysis. We then present eight design patterns, following a format consistent with existing pattern languages. Finally, we conclude with an open call for contribution to build upon the pattern language foundation.

2 Design Patterns and Pattern Languages

A pattern language is a set of interrelated design patterns that serve to solve reoccurring problems (Alexander et al. 1977). The pattern language approach proposed by Alexander et al. (1977) is well established in architecture. The design patterns follow a consistent structure—problem, discussion, and solution—which clearly articulate how and when the pattern applies. The work also separates patterns based on a category of scale including ‘towns’, ‘buildings’, and ‘construction’, starting broad and general before narrowing the focus on the specific. The nature of the patterns is also practical and prescriptive in terms of how they apply. An example of these patterns is 'Necklace of Community Projects' that discusses the idea of central town halls that are only useful to local communities for political discourse if surrounded by types of activities and businesses relevant to the town (Alexander et al. 1977).

'Liberating Voices' is a pattern language promoting equitable and effective civic engagement (Schuler 2008). Schuler (2008) applies a different approach to the grouping of patterns not by scale, but instead by where they fall on the axis of theory and practice, and by the relationship to different types of communities and groups. We draw on this approach by Schuler (2008), as our patterns involve a level of abstraction from the prescriptive approach of Alexander et al. (1977). Studies with gardeners have provided a mixture of themes relating to information and social interaction, physical environment, and use of resources. By employing a pattern language to support interaction designers, we aim to build on the understanding of existing work to guide the design of pervasive technology.

'Beautiful Trouble' (Boyd and Mitchell 2012) provides resources for activists, divided into modules that are networked similar to a pattern language. It consists of high-level categories including tactics, principles, theories, case studies, and practitioners. Each category comprises different entities—the equivalent of patterns—with relationships between entities. Beautiful Trouble provides a rich visualisation of the links between all entities, which is where we draw a parallel with pattern languages. As our proposed design patterns grow in number, we will explore ways of presenting the interconnectivity inherent in pattern languages.

In HCI, design patterns and pattern languages are used in a number of areas, as highlighted by Pan and Stolterman (2013) who reviewed and explored the current and future trajectory of pattern language use within research. They describe the variety of different uses, which include approaches to user interface design problems (Seffah 2010), programming patterns in object-oriented software development. Pattern language use in HCI also includes specific design contexts such as the domestic setting (Crabtree et al. 2002) and research fields such as computer-supported cooperative work (CSCW) (Martin et al. 2001).

Specific to the goals of environmental sustainability, Knowles et al. (2014) outline a set of design patterns of persuasion, using examples of eco-feedback technologies and understanding their impact on behaviour change. Examples of their patterns include the following: 'Designing to the value', which is concerned with personal values and how to address this in design by exploring underlying causes, and 'Consistency', which refers to the importance of the way persuasive messages to encourage behaviour should avoid introducing doubt. While behaviour change departs from our intent, a focus on sustainability via eco-feedback is relevant. This presents opportunities for further generalisation of similar studies in order to develop links between sets of design patterns.

Finally, Wania and Atwood (2009), while exploring real-world use of pattern languages by designers and in the ICT industry, suggest they hold value to facilitate communication between designers and stakeholders, more so than they exist to guide design. This traces back the original intention of the pattern language developed by Christopher Alexander (Wania and Atwood 2009).

2.1 The Problem of the Single Case

The choice of employing case study methodologies in areas usually guided by constructivism such as interaction design, but also action research, has prompted not only quantitative research proponents but also commentators within qualitative research domains to critically assess theory-building efforts. Gustavsen (2003) speaks of 'the problem of the single case' lamenting the unequal distribution of knowledge beyond the community, the lack of transferability of findings and interventions, and the difficulties in re-appropriating prior findings to new contexts (Foth 2006).

Scholars that question the transferability of qualitative research outcomes argue that the individual context found in a particular community, environment, or local setting renders the research results and findings idiographic. Martin et al. (2001) discuss the balance that must be struck between general design guidelines and those that are specific to a single case. However, just because a study's main objective is not to deliver nomothetic research insights, does not deny the fact that new knowledge is generated in rigorous case study research. Gustavsen (2003, pp. 162–163) concludes that

> To learn from practices, research needs to develop social relationships; internally within the research community as well as in relation to other actors. 'The new production of knowledge' as identified by Gibbons and colleagues (Gibbons et al. 1994) is above all a network activity, and research cannot stay outside this process and remain as isolated individuals looking at the world from up above.

The purpose of our design patterns has been to create guidelines that are useful and can be adapted to different types or scales of gardens. The outcome of design when applying each pattern will differ based on the contextual information of each case, and while these patterns could be applied to designing for city farms, grass roots communities, or residential gardeners as we have studied, they are not limited to these cases.

3 Urban Gardeners and Gardening Communities

A growing body of research explores the different needs and values of community gardens, both in Australia and internationally (Heitlinger et al. 2013; Odom 2010). Residential gardening on a household scale, typically undertaken by individuals, has been studied as 'pottering' activity by Taylor et al. (2008). These studies provide additional support for the development of our design patterns.

We conducted three studies between 2010 and 2014 with different gardening communities in Brisbane, Australia. We draw on the themes from these studies in order to develop our design patterns. These three types of gardening practices we studied represent a UA scales of macro (large shared location), meso (smaller community organising group events), and micro (individual practice).

3.1 City Farm

We conducted a study at Northey Street City Farm (Fig. 1, www.nscf.org.au), located 2 km north of Brisbane's CBD. The findings of this study were published by Lyle et al. (2013). Our study involved participant observation (through regular volunteering onsite) and five semi-structured interviews, all analysed through a coding and thematic analysis process that derived four key themes (Lyle et al. 2013). The findings gave insights into four themes.

- There is a preference for face-to-face communication, which has led to breakdowns in communication as a result of a flat hierarchy and a large physical site.
- Despite a number of income stream, NSCF is dependent on ongoing work of volunteers, demonstrating the need for frugal investment in any new technology, and to respect the commitments of volunteers.

Fig. 1 Northey Street City Farm, located to the north of Brisbane

- NSCF does not exist in a vacuum, and while participants are interested in the bigger picture of UA in Brisbane, those interviewed were content to focus their involvement on NSCF.
- The physical environment is naturally constrained, located on a floodplain (allowing such a large UA community so close to the city centre), and regular flooding stresses the operation and requires disaster management.

3.2 Grassroots Community

We conducted a study with Permablitz Brisbane (www.blitzbrisbane.org), a community that organises daylong events throughout Brisbane overhauling backyards (such as in Fig. 2) following permaculture design principles. The findings of this study were published by Lyle et al. (2014).

The group is the local chapter of the international Permablitz movement that began in Melbourne. Our study involved participant observation through volunteering at six *permablitzes*—events where local members of a community are invited to help transform a garden, typically a backyard, following a tailored permaculture design plan. We also conducted semi-structured interviews

Fig. 2 Site of a Permablitz event in Brisbane

with event organisers to develop an understanding of the process that goes on in planning and running events. The findings gave insights into four themes:

- Participants who organise the events have a high level of involvement, but also tend to burn out, which is attributable to the scale and scope of each event.
- Operating entirely with volunteers, the lack of financial resources limits their activities. Resourcefulness, however, is highly valued, and those involved borrow gardening tools and equipment wherever possible.
- It is important for the material resources and volunteers for events to be sourced from the local community, because after the event occurs, it will depend on the efforts of the local community to maintain the garden and serves to encourage interaction among community members.
- Permablitz Brisbane does not want to be a central authority on permablitz events and would rather just provide information and empower local communities to organise and run gardening events independently.

3.3 Residential Gardeners

Our study of residential gardeners in Brisbane included a survey (36 participants); a convergent semi-structured interview process (seven interviews, participants located in Brisbane and surrounding areas); and a focus group and semi-structured interviews with gardening experts (which involved an excursion to a nearby community garden, as shown below in Fig. 3). The experts were involved in a mix of managing community gardens, running educational workshops and hosting national television programmes about gardening, for over 10 years. Our findings gave insights into six themes:

- That experimenting and observational learning is a common approach to gaining gardening experience, rather than focusing on didactic learning.
- Gardening is often done as a recreational activity, and as such, it is often a low priority in the life of the urban gardener, so the time invested is limited.
- Gardeners discuss their success and failures, as well share their produce with family and friends, which serves to motivate ongoing engagement.
- Discerning the quality of online information can be intimidating, especially when navigating online opinions.
- Information sources are often specific to a particular location, and for participants in Australia, it is difficult to evaluate the relevance of gardening information for the Northern Hemisphere.
- The main areas where gardeners encountered difficulty were as follows: in understanding how they could make use of space; maintain a garden, making decisions about what/when to plant; and learning the practice of gardening and share their experiences.

Fig. 3 Expert gardeners inspecting a community garden near Queensland University of Technology in Brisbane

4 Design Patterns for Urban Gardening

The process of pattern creation is similar to that employed by Martin et al. (2001), which begins by exploring the themes and findings of prior work, looking for signs of a clear problem with a possible solution. We have—by exploring links and overlaps between findings of different studies—developed eight patterns.

As part of expressing the these design patterns, we are also interested in how they relate to patterns identified in similar fields, including Schuler (2008) and Knowles et al. (2014). For example, there is a link between the pattern we propose 'Understanding Impact', which can involve reflecting on gardening practice, and the pattern 'Retreat and Reflection' by Schuler (2008).

As part of understanding in pragmatic terms how different patterns can be used, examples are provided for existing technology or design efforts that show capabilities and limitations of each pattern. The structure we have used to explain our patterns follows a similar format to Schuler (2008). This approach is prescriptive, framing the design patterns on problems and their solutions, and although this ensures consistency, we recognise that this limits the types of patterns we can create (Martin et al. 2001):

1. **Name and Number**
2. **Problem**—statement of the reoccurring challenge;
3. **Context**—an overview of which types of gardening communities the issue applies;
4. **Discussion**—detail of where and how the problem occurs, examples of ways in which the problem has been approached by existing systems;
5. **Solution**—criteria or approaches that are likely to address the problem.

We have framed these patterns along two axes, the first covering themes of resources, information, and social connection; and the second highlighting whether the pattern applies to the physical or digital space. The patterns sharing axis are indicative of the links between patterns, as shown in Fig. 4.

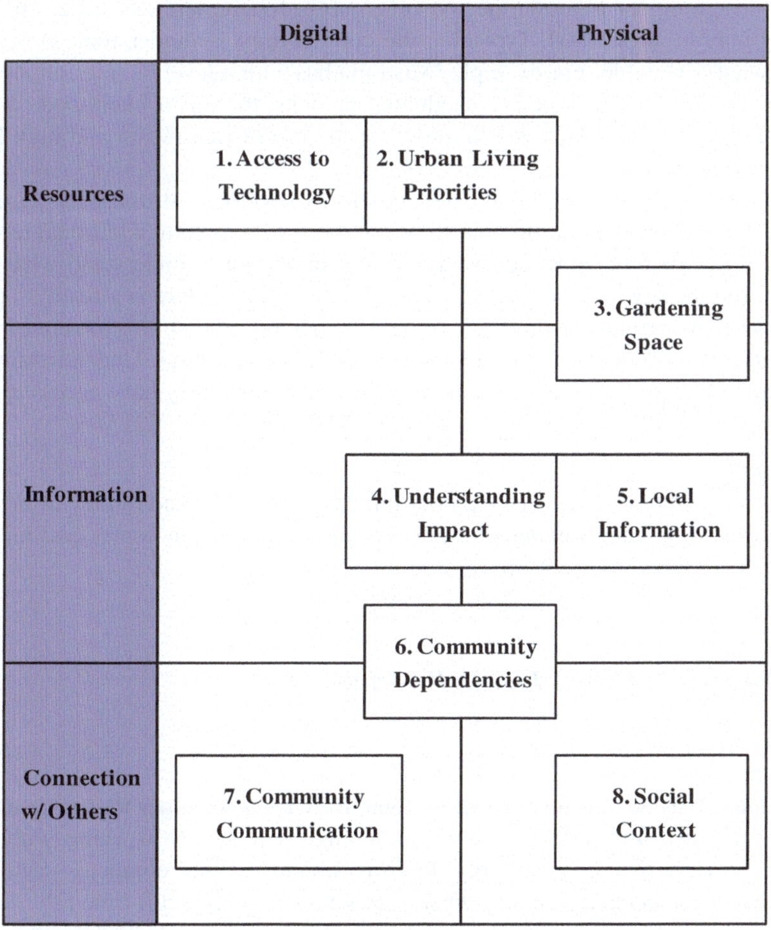

Fig. 4 Design patterns visualised in relation to each other

4.1 Pattern 1: Access to Technology

Problem
Gardeners can come from any background and as such have a wide variety of access to existing technology. Access to technology refers to whether an audience has a particular gadget or service, and their ability or willingness to use it as part of gardening practice.

Context
This problem applies to individuals and communities, whenever the intent is to design interactive technology. The context varies depending on the available resources of a community and the target demographic of design.

Discussion
When designing for a known person or group, infrastructure and access to technology may be prescribed. Typically, the context must be understood in order to know what is suitable. For example, Australia has a high level of smartphone market penetration, and if targeting residential gardens, there is a likelihood of high-speed Internet access. This would allow for the use of rich media and high levels of interconnectivity.

Communities on the other hand, such as Northey Street City Farm or Permablitz Brisbane, are limited in time and money to invest in additional technology or infrastructure. In these instances, it is important to understand what technology community members already use or what infrastructure is already in place, and how is it currently used. With this understanding, the ability to repurpose or make use of technology as part of a design will become clear. Understanding the role technology plays in the lives of gardeners, and when they have access to technology, will result in a more inclusive design (Heitlinger et al. 2013).

Solution
Designers need to consider the existing infrastructure; time and money to invest in new technology; and attitudes of gardeners to different technologies and incorporate these preferences accordingly.

4.2 Pattern 2: Urban Living Priorities

Problem
Gardening demands an investment of time, over time. Whether it is time used to engage in gardening, learning about gardening, or relates to working with other people in a gardening community, this investment of time competes with other priorities in urban environments, where gardening is rarely a full-time job.

Context

This pattern can be used with any type of gardening practice where competing interests, priorities, and time constraints may exist. Residential gardeners may only have irregular time to invest and treat food growing as a hobby, undeterred by failure with financial security and access to food at shops to respond to lost crops. Community gardens often involve the time and resources of volunteers, in addition to tight budgetary constraints, and this may influence crop choice decisions, based on what is easiest to grow for the largest volume or financial return.

Discussion

This pattern indicates a need to understand the priorities and values of the people and communities that are the demographic of design. For urban residents, the level of commitment they have available to engage in gardening, either individually in their own space, or as part of a larger community, will depend on their work-life balance. For many, gardening will be a hobby (Taylor et al. 2008), and the frequency and duration of time available to garden will be limited, creating a situation where plants may not receive the attention required to yield a significant result. However, the effectiveness and efficiency of growing food may be less important than the act of participation (Heitlinger et al. 2013).

In two studies of gardening communities at a city farm and grass roots movement, there was a dependence on volunteers and a lack of resources to invest in time or infrastructure. In the case of Permablitz Brisbane, the clash between involvement as an organiser and other priorities resulted in volunteers halting or reducing their involvement.

While resources are tight as part of the daily operations of these communities, the idea of resourcefulness is an important value, an aspect shared by other community gardening groups (Odom 2010).

For technically competent, but time-poor gardeners, systems that provide automation are an example of responding to this pattern. An example of existing technology that responds to this pattern would be *Flower Power* (www.parrot.com/usa/products/flower-power) that uses sensors to make information available in real time, wirelessly to smartphones. These data help reduce or streamline the time needed to invest in the garden. *Hydrawise* (hydrawise.com) is a smartphone application that uses local weather information to assist with the management of water and automated irrigation.

Solution

Gardeners need help to enjoy their gardening time and manage its effectiveness. The approach of designers should involve understanding where gardening sits among the gardener's priorities. Ensuring that barriers to entry remain low, and leveraging context-aware pervasive technology, is one approach that would address this problem.

4.3 Pattern 3: Gardening Space

Problem
The ability to have a successful crop with specific plants and the level of required maintenance will be dictated by the physical gardening context. How do gardeners know what is possible, what is ideal, and what hazards exist within a given space?

Context
This pattern applies to all types of gardens and gardening communities as it focuses on the physical environment in which gardening occurs, and the planning, opportunities, and limitations of this space.

Discussion
It is difficult to prescribe if-then statements to explain how different spaces impact gardening, however, how a gardener gardens will differ between a pot plant on a balcony, a surburban backyard garden, a community garden or a city farm. The objectives of each type of garden also shape the best options, from the need for low-maintenance gardening for a busy urbanite, to the need for high yield if selling/trading produce at markets.

This pattern covers social attributes of how gardeners interact with the space and physical attributes of the natural environment. While limited irrigation might impact gardens, a more serious example of physical limitations is natural disasters, e.g. floods and bush fires. Considering disaster management as part of interacting with the physical environment can have an impact on gardening safety.

While drafting paper and sketching tools are useful (and drafting on paper is how the Brisbane Permablitz plans are presented), there is opportunity to build on computer-aided design tools, such as Google SketchUp, 3D Studio Max, or Blender, with the restrictions and knowledge of the physical environment, in an accessible way to assist gardeners and communities to understand their available space.

Solution
The scale of the gardening space and the environmental conditions of the space must be known in order to understand what meaningful options or approaches to gardening should be considered. Designs should enable gardeners to realise the potential of their space and plan around the limitations.

4.4 Pattern 4: Understanding Impact

Problem
Gardeners need to be able to understand the practice of gardening, both in individually and in communities. Having or gaining the knowledge of what options

there are when planting, harvesting, or treating pests/diseases is not always known, and the non-deterministic nature of gardening outcomes is problematic without a frame of reference.

Context

This pattern applies to all scales of community, but in different ways. A residential gardener with limited time and a small gardening space will have low amounts of gardening activity, and the ability to recall the changes in the garden over time will be relatively easy. Whereas for a community garden, a large amount of work performed by multiple people may lead to difficulty in understanding the impact.

Discussion

Decisions made while gardening either individually or as part of a group create expectations—planting seeds lead to growth and harvest. A number of factors will impact the outcome of this process, some of which are controllable, such as the ongoing garden maintenance, weeding, pest, and disease management. Others, such as weather, can only be planned for based on previous trends. In situations where crops fail, this failure can be demotivating. Reflecting on the process, to establish an understanding of why the outcome was a certain way, can help improve knowledge and experience.

Experimenting and observing the impact of gardening decisions is a method by which individual gardeners learn—this is a theme of our interactions with residential gardeners. Survey participants were asked how they gain a sense of seasonality and the most common response related to the process of experimentation, and observation of their own experiments, local farmers markets, and available supermarket produce.

For communities, there is value in understanding the impact of their efforts, as well as to understand their efforts in the context of other community gardening in the surrounding area. For Permablitz Brisbane, understanding the impact of each event on the garden is a means of celebrating the work of the volunteers, to encourage ongoing engagement and reduce risks of burnout (Lyle et al. 2014). Permablitz Brisbane and Northey Street City Farm benefit from understanding the wider context of UA in Brisbane (Lyle et al. 2013, 2014).

This pattern resonates with Liberating Voices pattern #136 'Retreat and Reflection' (Schuler 2008), which discusses the importance of reflection, but in the context of taking a step back from the pressures of modern life. This is relevant to *Understanding Impact* in cases where community gardening involves volunteers. Retreating and reflecting can create space for volunteers to appreciate the effort and impact they have invested.

This pattern also corresponds with 'Facilitate reflection' (Knowles et al. 2014), which calls for celebration and reinforcement of values, in order to avoid being overwhelmed when faced with how to respond to environmental sustainability. An example of how this is relevant to *Understanding Impact* is in encouraging enthusiasm in volunteers, celebrating progress to avoid burnout.

Examples include a journal of gardening activity (online or offline) or a regularly updated newsfeed on a community garden Website. Permablitz Brisbane's

Website (www.blitzbrisbane.org) includes a written article with images of each event, details about what occurred, and follow-up events.

Solution
Gardeners need to be able to understand the impact their efforts have in the garden as part of their own learning and to encourage ongoing engagement. Designers should explore ways to create opportunities for gardeners to appreciate their own gardening and the gardening of their community.

4.5 Pattern 5: Local Information

Problem
Information about gardening practice is readily available via online resources and in books, but its relevance is tied to specific locations that may differ from the gardener's own. The location context of the information is not always explicit or accurate.

Context
Individual urban gardeners need to understand their gardening context. For gardening communities, the physical size of the gardening space may be greater; however, the shared knowledge of the community can contribute to a more accurate overall set of aims and ideas in the garden. It may do so at the expense of social cohesion if strongly held conflicting opinions and experiences exist.

Discussion
For events such as those organised by Permablitz Brisbane, sourcing locally and temporally specific information is crucial to understanding the surrounding environment, the available suppliers for materials, and what can be planted, according to the season. Residential gardeners benefit from this information, and similarly struggle, when they search online and stumble across conflicting suggestions for how to respond to a pest. Our study of residential gardeners in Brisbane found that information sourced online was often implicitly or explicitly assuming USA or European as the users' location, limiting its usefulness.

ABC Gardening Australia is one of the most popular responses on our survey of residential gardeners (mentioned by seven respondents) regarding where they source information. They regularly provide information through a national television programme, Website, and social media presence. Their Web presence includes a 'vegie guide,' (www.abc.net.au/gardening/vegieguide)—a planting calendar that provides different recommendations depending on the selection of one of five climatic zones. This is an example of tailoring information to local context, although depending on the level of detail, dividing a country the size of Australia into five regions may be too coarse. Despite this, a number of residential gardeners still depend on the ABC Gardening Australia as an authoritative source.

Another example is the exploration of sensor and automation systems that take into account different elements of your context and provides information and advice on garden maintenance that is tailored, such as Smart Garden Watering (www2.smartgardenwatering.org.au).

Solution
Gardening occurs in the real physical world, and as such understanding the role of location, season on macro-, meso-, and microlevels is critical for information to be useful. Metadata should encode the temporal and spatial context in order to be effective. Training can provide a better understanding on how to use this contextual information and to be able to filter by relevance.

4.6 Pattern 6: Community Dependencies

Problem
Gardeners transition in and out of involvement of gardening communities, and with this, they both bring and take their knowledge and experience. This can have a negative impact on a community's autonomy where the involvement of certain members who represent key tacit knowledge creates a dependency. When key members leave, move on, or become less involved, how will communities continue?

Context
This pattern applies to gardening communities, online and offline, and the ways in which their individual experiences are shared and stored. The impact differs between communities of residential gardeners, such as a gardening club, compared to a community with shared involvement in a specific garden.

Discussion
We found, through our study with Permablitz Brisbane, that while the organisers were interested in helping provide information on running events (such as manuals for event hosts, the organisers and planners etc.), there was a concern that local communities may become dependent on the expertise and experience of Permablitz Brisbane, and look to them as an authority.

While there is value in recognising the experience and expertise of individuals and groups, local communities should be empowered to develop their own rather than rely on others. Encouraging documentation of systems and decentralising the way groups interact reduces the impact of tacit knowledge loss when people move on. Storytelling systems may also encourage oral mechanisms of transferring tacit knowledge to others.

Wiki systems provide an example of simple methods of capturing and updating information about different processes, practices, and expertise of a group. Wikis are not without issues, however, as a lack of maintenance leads to the relevance of the information becoming stale over time.

Solution
The solution to community autonomy is to both encourage community members with useful information and practices and discourage dependence on centralised systems of control. Approaches worthy of consideration include designing for wide dissemination of both tacit and codified knowledge about specific communities and general gardening and empowering individuals to contribute using open-source and/or collaborative systems.

4.7 Pattern 7: Community Communication

Problem
When gardening at a shared location, members of a community may prefer face-to-face communication; however, other gardening communities exist solely online where interactions are asynchronous. The means of communication create opportunities for communication breakdown, and designers should carefully consider what means of communication is appropriate.

Context
Understanding the way gardeners communicate, and whom they communicate with, is crucial. Community members who are co-located may explicitly prefer face-to-face communication. Individual attitudes towards different technology or means of communication also provide scope to shape designs.

Discussion
Our own study with NSCF found a distinct preference for face-to-face interactions, which across the large site created potential situations for communication to break down. This is related to their flat hierarchical structure and the locations of the different community members. During the interviews at NSCF, there was also a distrust of social media technologies and an interest in a closed network of communication. For residential gardeners, and Permablitz Brisbane, the use of social media was more prevalent and accepted.

This pattern is similar and overlaps with Liberating Voices patterns #251 'Online Deliberation' and #260 'Community Networks' (Schuler 2008). The first is directly applicable to our case of communication breakdown, by encouraging the design of technology that allows for collaboration when the gardeners onsite are physically separated. One way that NSCF had engaged with members who were not co-located was to use Google Docs for sharing and collaboration. Pattern #260 relates to communities engaging with different technology and the patience required to convey the benefits.

Designs that respond to this pattern could explore synchronous or asynchronous communication, depending on how the communities already engage with each other. For Permablitz Brisbane, who are not co-located, mailing lists and CMS have been used as the means of communication, and therefore, future design for their case might explore whether a synchronous communication channel could complement their existing approach.

Solution

Do not expect gardeners to change the way they communicate, instead design to complement or repurpose their technology and communication preferences to allow for ease of adoption. Alternatively, design with flexibility of communication explores different types of media and interfaces for communication.

4.8 Pattern 8: Social Context

Problem

Motivation to start and maintain a garden often involves social interaction. How can design take advantage of this opportunity for gardeners with established social support and encourage gardeners with limited support?

Context

This pattern relates to community and individual gardening. In the case of individual gardening, 'individual' is perhaps a misnomer, as it does not consider the social support from family and friends. For communities, the social interactions will include a mix of friends and acquaintances, and the interactions will be focused more on the efforts of the group rather than individual reflection.

Discussion

Our study with residential gardeners revealed that the role of family and friends as part of sharing experiences about gardening was important to most participants and at least partially responsible for their continued engagement. For communities such as NSCF, understanding the importance of communication between members is important for social cohesion and is necessary for effective gardening in a shared space.

The social interaction at events organised by Permablitz Brisbane involves an intense time of socialisation (and gardening), but outside of the event, the interactions are much less frequent and intense. Understanding the social support and interactions that take place, and the frequency and intensity of them will allow more tailored approaches in design.

Where #7 'Community Communication' above is concerned with the digital means and preferences for communication, this pattern is focused on the social groups and context in which gardening takes place—Who do gardeners engage with while gardening, for motivation, for information? Understanding that these social networks can be part of encouraging gardening practice means they can be considered as part of the design.

Solution

As a motivation for initial and ongoing engagement in gardening practice, individuals and communities benefit from communicating with friends and family, or other gardening communities about their practice and gardening experiences. Designers should look to integrate social networks and social connectivity into the experience of gardening.

5 Conclusions

Our contribution has been to outline eight initial design patterns that can be applied in the context of different types of urban gardening. The development of these design patterns is based on three studies of different types of UA, a review of the literature, and other pattern languages. The design patterns we present are a foundation on which a pattern language can be developed, and we encourage others to contribute, both to refine and enrich the proposed patterns and to add new patterns. This is a foundation for a niche area of design, and there is opportunity to adjust the scope of the language in different directions, including for other types of UA, or to encompass more of the sustainable HCI research field.

The proposed design patterns, while developed from the outcomes of existing studies, require further testing to validate, iterate, and refine. We propose future development of a pattern language through the design of interactive and pervasive systems that aim to improve the way in which gardeners engage with each other, reflecting on their individual and community contributions to food production in urban spaces. Future work could also explore if a metalanguage could, at a level of abstraction from this foundation, create tighter links between languages—e.g. where there is overlap with ecological sustainability, (Knowles et al. 2014), civic engagement (Schuler 2008), and food systems (Hearn et al. 2014). Drawing inspiration from the visualisation of *Beautiful Trouble,* (http://explore.beautifultrouble.org) there is an opportunity to increase the utility and highlight the connectedness of these design patterns through interactive presentations.

Acknowledgements This research study was supported under the Australian Research Council's Linkage Projects funding scheme (project number LP100100232). Dr. Jaz Choi was the recipient of an ARC Australian Postdoctoral Fellowship (Industry), and Dr. Peter Lyle was the recipient of an Australian Postgraduate Award (Industry).

References

Alexander, C., Ishikawa, S., Silverstein, M., Jacobson, M., King, I. F., & Angel, S. (1977). *A pattern language: Towns, buildings, construction*. Oxford: Oxford University Press.

Bellows, A. C, Brown, K., & Smit, J. (2003). Health benefits of urban agriculture. *Community Food*.

Boyd, A., & Mitchell, D. O. (2012). *Beautiful trouble: A toolbox for revolution*. New York: OR Books.

Crabtree, A., Hemmings, T., & Rodden, T. (2002). *Pattern-based support for interactive design in domestic settings*. Paper Presented at the Proceedings of the 4th Conference on Designing Interactive Systems: Processes, Practices, Methods, and Techniques, London, England.

Foth, M. (2006). Network action research. *Action Research, 4*(2), 205–226.

Foth, M., Choi, J. H., & Satchell, C. (2011). Urban informatics. In *Proceedings of the ACM 2011 Conference on Computer Supported Cooperative Work (CSCW '11)* (pp. 1–8). New York, NY, USA: ACM. doi:http://dx.doi.org/10.1145/1958824.1958826

Gibbons, M., Limoges, C., Nowotny, H., Schwartzmann, S., Scott, P., & Trow, M. (1994). *The new production of knowledge: The dynamics of science and research in contemporary societies*. London: Sage.

Gustavsen, B. (2003). Action research and the problem of the single case. *Concepts and Transformation, 8*(1), 93–99.

Hearn, G., Collie, N., Lyle, P., Choi, J. H., & Foth, M. (2014). Using communicative ecology theory to scope the emerging role of social media in the evolution of urban food systems. *Futures, 62*, 202–212.

Heitlinger, S., Bryan-Kinns, N., & Jefferies, J. (2013). *Sustainable HCI for grassroots urban food-growing communities*. Paper Presented at the Proceedings of the 25th Australian Computer–Human Interaction Conference: Augmentation, Application, Innovation, Collaboration.

Knowles, B., Blair, L., Walker, S., Coulton, P., Thomas, L., & Mullagh, L. (2014). *Patterns of persuasion for sustainability*. Paper Presented at the Proceedings of the 2014 Conference on Designing Interactive Systems, Vancouver, BC, Canada.

Kortright, R., & Wakefield, S. (2011). Edible backyards: A qualitative study of household food growing and its contributions to food security. *Agriculture and Human Values, 28*(1), 39–53. doi:10.1007/s10460-009-9254-1.

Lyle, P., Choi, J. H., & Foth, M. (2013). *HCI for city farms: Design challenges and opportunities*. In Proceedings of the 14th IFIP TC13 Conference on Human-Computer Interaction (INTERACT) (Lecture Notes in Computer Science No. 8120, pp. 109–116, Sep 2–6). Heidelberg, Germany; Springer. doi: 10.1007/978-3-642-40498-6_7.

Lyle, P., Choi, J. H., & Foth, M. (2014). Designing for grassroots food production: an event-based urban agriculture community. In Proceedings of the 26th Australian Computer-Human Interaction Conference on Designing Futures: The Future of Design (OzCHI '14) (pp. 362–365). ACM, New York, NY, USA. doi:10.1145/2686612.2686666.

Martin, D., Rodden, T., Rouncefield, M., Sommerville, I., & Viller, S. (2001). *Finding patterns in the fieldwork*. Paper Presented at the Proceedings of the Seventh Conference on European Conference on Computer Supported Cooperative Work, Bonn, Germany.

Odom, W. (2010). *Mate, we don't need a chip to tell us the soil's dry: Opportunities for designing interactive systems to support urban food production*. Paper Presented at the Proceedings of the 8th ACM Conference on Designing Interactive Systems, Aarhus, Denmark.

Pan, Y., & Stolterman, E. (2013). *Pattern language and HCI: Expectations and experiences*. Paper Presented at the CHI'13 Extended Abstracts on Human Factors in Computing Systems, Paris, France.

Schnell, S. M. (2013). Food miles, local eating, and community supported agriculture: putting local food in its place. *Agriculture and Human Values*, 1–14. doi:10.1007/s10460-013-9436-8.

Schuler, D. (2008). *Liberating voices: A pattern language for communication revolution*. Cambridge: MIT Press.

Seffah, A. (2010). *The evolution of design patterns in HCI: From pattern languages to pattern-oriented design*. Paper Presented at the Proceedings of the 1st International Workshop on Pattern-Driven Engineering of Interactive Computing Systems, Berlin, Germany.

Taylor, A. S., Wyche, S. P., & Kaye, J. J. (2008). *Pottering by design*. Paper Presented at the Proceedings of the 5th Nordic Conference on Human–Computer Interaction: Building Bridges, Lund, Sweden.

United Nations. (2014). *World urbanisation prospects: The 2014 revision*. New York: UN Department of Economic and Social Affairs (Population Division). Retrieved from http://esa.un.org/unpd/wup/index.htm.

Wania, C. E., & Atwood, M. E. (2009). *Pattern languages in the wild: Exploring pattern languages in the laboratory and in the real world*. Paper Presented at the Proceedings of the 4th International Conference on Design Science Research in Information Systems and Technology, Philadelphia, Pennsylvania.

Part II
Citizen Action

ICTs and Technical Agency: A Case Study of a Rural Brazilian Community

Lorena Melgaço and Katharine S. Willis

Abstract This paper explores the role of information and communication technologies (ICTs) in rural, marginalised communities. We explore Feenberg's concept of technical agency in order to understand the ways in which technological capacity might contribute to community development. We discuss three conditions of technical agency as outlined by Feenberg: power, knowledge and appropriate occasion. We consider how this framework might enable an approach to understanding the role of ICTS in the particular socio–spatial contexts in marginalised communities. In order to test Feenberg's approach, we discuss a case study of a village in rural Brazil: Noiva do Cordeiro. The community is recognised regionally as a pioneer in the way that it has mobilised a range of ICTs for the benefit of the community. We will conclude by discussing how ICTs can reinforce existing or introduce technical agency in a Feenberg framework and how this might contribute to community development.

Keywords Technical agency · Community development · Community informatics · Marginalised communities · Knowledge networks

1 Introduction

The correlation between technological advancement and urban development has been discussed by a number of theorists in terms of the physical, social, economic and political factors (de Sola Pool 1990; Sassen 1991; Graham and Marvin 2001;

L. Melgaço (✉) · K.S. Willis
School of Architecture, Design and Environment, Plymouth University, Plymouth, UK
e-mail: lorena.melgacosilvamarques@plymouth.ac.uk

K.S. Willis
e-mail: katharine.willis@plymouth.ac.uk

M. Foth et al. (eds.), *Citizen's Right to the Digital City*,
DOI 10.1007/978-981-287-919-6_6

Amin 2002; Graham 2004; Massey 2004; Swyngedouw 2004; Castells 2011). Technology and urban development are often perceived as complimentary, and as such technological frameworks and social structures are often understood as interlinked. According to Graham and Marvin (2001), technological networks are systems that help to shape the urban environment since 'as capital that is literally 'sunk' and embedded within and between the fabric of cities, they represent long-term accumulations of finance, technology, know-how, and organisational and geopolitical power, (2001: 12). Latham and Sassen (2005) introduce the concept of socio-digital formations to address this intertwining of social and digital structures. These are 'electronic information and communication infrastructures resulting from various mixes of computer-centred technical capabilities and the broad range of social contexts that provide the utility logics, substantive rationalities and cultural meanings for the particular types of digital interaction involved' (Sassen 2006: 208). Socio-digital formations are shaped by social, political, economic and even visual structures, which help define 'novel social forms' through the juxtaposition of new forms of organisation, interaction and electronic space (Latham and Sassen 2005: 2).

Yet, despite the focus on how ICTs contribute to urban transformation, there is a less understanding of how ICTs are embedded within the particular socio-cultural contexts of marginalised communities. The growing field of community informatics (CI), situated in the intersection between the social and the technological, represents the increasing recognition of the role of technology in enabling and empowering community processes (Gurstein 2007: 11), particularly in developing world contexts. Within CI, there is a general understanding of the political character of technology, because 'sociologies of connectivity are predicated upon the architecture of networks, since the architecture governs positionalities and relationships, determining who has power, who can exercise control, and who is included' (IT for change 2014: 31).

The problem we seek to address is how ICTs can benefit and empower communities that are marginalised. Here we use the term marginalisation in the sense that it describes 'a form of acute and persistent disadvantage rooted in underlying social inequalities' (Unesco 2010). According to Demo such marginalisation contributes to 'the inability of a given community and its individuals to mobilise within the various spheres of individual and communal life' (Demo 1994). This includes communities that are in some way socially, economically or geographically excluded from mainstream society. Marginalised communities are often recognised as those that could benefit most from access to ICTs in order to address issues of geographic and economic exclusion (Unwin 2009; IFAD 2003). However, one of the underlying problems of the role of ICTs in community development is that increasing technological access to ICT networks, such as those characterised by Castells (2007), actually contributes to the perpetuation of existing divides. This is because access does not in itself overcome the broader challenges of the lack of technical skills, poor economic opportunities and existing democratic divides (Mossberger et al. 2003). ICTs can never be a simple counter to marginalisation because even when technology is made available to marginalised groups; 'what usually happens is inclusion in the margins; they may be a little but more included, but they continue to the marginalised' (Demo 2007: 06). In fact in many cases, the introduction of ICTs in marginalised communities leads to the reproduction and reinforcement of existing social relations and power relations (Mbarathi and Diga 2014). That is because the communities involved often lack the circumstances and the understanding of the importance of technology for empowering themselves either as individuals or as a collective: they lack agency.

This can be seen in many of the current strategies that seek to overcome digital exclusion in developing countries. A large-scale example is the One Laptop per Child (OLPC) project.[1] The OLPC project, championed in the mid-nineties, aimed to develop low cost, resistant, open-source computers in order to 'empower the world's poorest children through education'.[2] Different physical constraints were taken into account when the piece of hardware was developed; nevertheless, the project still neglected many of the social, cultural and political dimensions associated with the contexts it was to be deployed into (Willoughby 1990). For example despite being open-source and low cost, the project still struggled to provide access to technology, because the lack lies in the access itself (Souza 2009). The OLPC project demonstrates that technology provision alone is not sufficient to overcome the socio-cultural divides that underpin and reinforce the lack of access to digital technologies (Mossberger et al. 2003; Baltazar 2009; Melgaço and Baltazar 2011).

In the next section, we discuss the role of ICTs and technical agency to support and develop collective agency in order to counter economic, social and digital exclusion. We first discuss the political dimension of technologies and the implications and outline Feenberg's concept of technical agency. The three conditions for Feenberg's technical agency are discussed: power, knowledge and appropriate occasion. This is contextualised in a case study, where we study a rural community in Brazil, Noiva do Cordeiro. The village was chosen because, although it might be characterised as marginalised, it has adapted ICT use to address the needs, skills and specific socio-spatial context of the community. We explore how ICTs have affected the community in terms of its everyday practices, and the consequent implications of this for an understanding of their technical agency. In the summary, we draw together the theoretical framework and the findings from the case study to reflect more generally on ICTs in marginalised communities beyond the Brazilian context.

2 Technical Agency

The issue of what constitutes 'access' to ICTs is often dominated by a narrative around ICTs context that focuses on 'how' technology can be used and not on 'why' its use should be developed with regard to its helpfulness to people's

[1] http://one.laptop.org/.

[2] http://one.laptop.org/about/mission.

everyday lives (Flusser n.d; Thackara 2001). The focus on a 'how' rather than a 'why' underlines a technologically deterministic approach in terms of the technology and the impact on people's lives. According to (Feenberg 1999: 77–78), 'determinism claims that technologies have an autonomous functional logic that can be explained without any reference to society' that is based on an approach to technology that sees it as universal, decontextualised and self-generating (Feenberg 2009). He argues that technological development needs instead to be understood as a social process, a site of social struggle or a Latourian 'Parliament of the things' (1993) on which political alternatives contend. Feenberg's approach is to argue for an understanding of the political dimension of technology and thus to consider the concept of 'technical agency' or 'technical citizenship' (Feenberg 2011). In this context, Feenberg argues that 'either politics becomes another branch of technology, or technology is recognised as political' (1999: 02), which he defines as political action within the technical domain (2011). This acknowledgement of the political dimension of technology means that it must be defined contextually and locally by the particular technology/society relationship and that technology can never be removed from a context, and therefore can never be neutral (1999: 213). In this chapter, we work with Feenberg's theories around the democratisation of technology and in particular his concept of technical agency.

The political nature of technology is particularly important to understand within the context of the challenges faced by marginalised communities. It is important to note that physical availability of computer and Internet does not correspond with access, as discussed in the OLPC project above, since access requires skills, opportunity and 'technical agency' to mobilise these skills within a given context. According to Gurstein, rather than responding to the 'digital divide' by extending opportunities for technology or Internet access, the approach should instead focus on 'finding ways of using ICTs to respond to issues and requirements that are meaningful and significant in the daily lives of individual users within their communities' (Gurstein 2003: 34). Whilst access to ICTs opens up 'the number of discourses and subject positions to which the individual becomes exposed, as well as by multiplying the participation forms available at the individual's fingertips' (Bakardjieva 2009: 94), Feenberg argues that it is only socially compelling when those individuals who use it have the agency or the 'capacity to act' on it, and most importantly, beyond it (Feenberg 2011). This considers technology and its various applications based on local knowledge and actions that address local problems when top-down strategies fail (Feenberg 2011). According to Feenberg, technical agency depends on three conditions: knowledge, power and an appropriate occasion (Feenberg 2011), which we expand on below.

2.1 Knowledge

Feenberg identifies two sorts of knowledge: that related to technical issues which he defines as a 'specialised knowledge' and that which comes from experience, which he calls a 'knowledge from below'. In the technical domain, these two forms of knowledge usually conflict, since specialists tend to overlook the needs of those affected by the technology (Feenberg 2011).

2.2 Power

Feenberg (2011) describes power in relation to how people can work with technologies in a way that is appropriate to their own needs. Through such appropriation, they also empower themselves and their community. According to Feenberg:

> involvement with a technology makes certain interests salient […] Once enrolled in a network individuals are motivated to address its failings and in some cases they also acquire potential power over its development. And the power of individuals within a network is quite different from that of individuals who have no connection to it. Because they are on the inside they can identify vulnerabilities and bring pressure to bear. (Feenberg 2011: 05).
>
> In Feenberg's thinking, power instead equates with being empowered.

2.3 Appropriate Occasion

Appropriate occasion is seen as circumstances in which the application of knowledge and power makes sense (Feenberg 2011), and refers to favourable circumstance in which knowledge and empowered citizens may enact social transformation. Appropriate occasion, in this sense, means not only the point at which technical agency is enabled, but also the circumstances that allow for community members to appropriate ICTs for their own benefit.

In the next section, we introduce a case study of particular marginalised community in rural Brazil, and in the discussion, we will work with these three conditions to explore the political dimension of technology within the everyday local practices of the community.

3 A Case Study of a Rural Brazilian Community: Noiva do Cordeiro

Noiva do Cordeiro (Noiva) is a rural community of around three hundred people on the fringes of Belo Vale, situated eighty kilometres from Belo Horizonte, the capital of Minas Gerais, Brazil (see Figs. 1 and 2). It can be characterised as marginalised since it can be considered to experience economic, social and spatial exclusion. Economically, it is a subsistence community with basic farming as a source of income, and for many years in the last century, the village experienced food poverty. The community is geographically isolated, as the closest town,

Fig. 1 Noiva do Cordeiro: 'a big family' (*source* https://www.facebook.com/noivadocordeirocomunidaderural/photos/pb.190044884409106.-2207520000.1436820022./548032818610309/?type=3&theater with permission)

Fig. 2 Noiva do Cordeiro community (*source* author)

Fig. 3 Noiva do Cordeiro and the collective and meeting spaces (*source* Based on Zerlotini 2014)

located sixteen kilometres away, can only be reached by a dirt road. There is only very limited public transport.

Historically, the village was socially isolated for a long period of time. The settlement formed in the 1890s when a local family was excommunicated from the local catholic community. Subsequently, the founders of the community of Noiva do Cordeiro established a settlement on the family's farmland and were forced to be entirely self-sufficient for over a hundred years. The community started to re-establish connections with the wider community only in the last decade of the twentieth century. As a result of this extended period of social isolation, Noiva developed a fairly unique social dynamic, where the village operates as a collective.

In the community, there is very little social hierarchy, and work is organised and managed collectively. All the resources, work and spaces of the community are shared, from childcare and clothes to housing and farming activities. Decision-making is also collective, and any decisions are made through meetings with all the inhabitants present, as 'only a few cannot decide for all' (Noiva's inhabitant 2013). The community ethos is based on the approach of 'all for one and one for all' (in Portuguese: uns para os outros) in the sense that people do not think and act individually, but collectively. The community relies primarily on subsistence agriculture and some tourism, as well as some more recent entrepreneurial initiatives. Everyday life is primarily organised by the women of the community, since almost all the men commute to work in nearby cities and are absent during the week.

The collective nature of the community extends to the spatial organisation of both housing and communal activities (see Fig. 3 for list of buildings and functions). The village has around twenty housing units mostly in shared living spaces, with residents living either in three collectives houses (the 'Mother house', the 'Yellow house' and the 'Big house' where a nursery is also located) or in other extended family units. The construction of new units is discussed in group meetings, and final consent is then given by D. Delina, the community matriarch (Zerlotini 2014). The 'Mother house' is a central space in the community and is both home to the village matriarch and several other inhabitants as well as being a more general meeting space. The collective division of the household work has led to the creation of equally collective spaces for laundry, cooking, babysitting as well as for group meetings, cultural activities and sports. These meeting spaces include a telecentre (located in the 'Big House'), a community centre where most activities and meetings take place, and a communal kitchen (see Fig. 3). These locations also support a range of social activities, including theatre, dance and computer classes, that are all organised by members of the community in their own time.

3.1 Case Study Method

The community was studied longitudinally through the years of 2012 and 2013 (Baltazar 2013; Kapp 2013; Melgaço 2013a, b; Zerlotini 2014). An ethnographic approach was taken, based on different field visits and interviews with a number of members of the community. The community is careful to control how it is presented externally and does not allow researchers to digitally record interviews.[3] Therefore, all references to quotes are derived from the researchers' written record of the conversation.

4 Results: ICTs in the Community

Partly as a result of the social isolation, but also due to the small size and isolated geographical location of the village, Noiva had very limited communication infrastructure or access beyond the community until 2005. The only form of communication infrastructure in place was basic landline access, with no mobile phone or Internet access. According to a village inhabitant

> Until 2005, we had no contact with anything. We were an isolated community in the world (Fernandes 2014).

[3]By doing so, they expect to avoid misunderstandings. As one inhabitant explains, 'after being recorded, there is no way words can be undone'. The community has recently experienced problems with the way they were portrayed in international newspaper articles. In 2009 Marie Claire Brazil (Cerenza 2009) ran a story about the village and that was picked up on by the UK press in 2014 (Bowater 2014; Roper 2014) with a series of sensational stories that characterised Noiva as a 'Brazilian all-woman village desperate for men'. In only a few hours, this story was reproduced by other worldwide newspaper and websites. Many even appropriated photos taken out of its professional context showing the women in provocative outfits, thus presenting an entirely false view of the community.

Around this time, the community actively set up a number of initiatives that would counter this marginalisation, primarily on the basis of ensuring the economic viability of the community. Overcoming the social isolation they had experienced up until the previous decade was recognised as a key goal by the community, since:

> The fact that we lived isolated harmed our communication with outsiders. The community had difficulty interacting and experienced economic and educational problems (Fernandes 2014).

This included the installation of a range of ICT infrastructures, establishing communications networks with the world beyond the village and finally building digital skills through training and access to computers. Since it was the first rural community in the state of Minas Gerais to have this infrastructure, the community became widely known as 'rural pioneers' in the field of ICTs. Below, we outline the initiatives undertaken under three categories: ICT infrastructures, communication with and beyond the community, and finally ICT skills and training. These three different approaches combined have not only ensured ICT access within the community, but have led to a technical agency that supports and develops the economic and social life of the community.

4.1 ICT Infrastructure

In 2006, the community successfully set up a telecentre through a pioneering partnership between the government, the village community association and CDI (Committee for Democratisation of Technology), a non-profit organisation that provides ICTs to underprivileged communities. This was the first such computer school in the State of Minas Gerais. The telecentre is called the Center for Inclusion and Development of Community Education (CIDEC) and was used as a pilot project, and similar telecentres were built in other towns in the region.

Some inhabitants were initially given training by the CDI, but the knowledge was soon shared within the community. Initially, they used the newly introduced technology to collectively plan, manage and distribute the supplies from the subsistence farming. The telecentre is a room equipped with eight computers (Fig. 4). As with other collective spaces in the community, it is collectively run and people voluntarily help one another with technical problems and training.

An inhabitant recalls the training sessions given to people when the Internet was installed. During this period, people had to learn to use the computer. Élida remembers she had never seen a computer before and then:

> suddenly it was there, right in front of me.

Although the community negotiated the introduction of the telecentre, it did not have Internet access. In 2008, the community worked together with the Vale Foundation,[4] a national organisation that works on social, economic and environmental empowerment projects to install the Internet for a year in the community.

[4]http://www.fundacaovale.org/en-us/a-fundacao-vale/Pages/default.aspx.

Fig. 4 CIDEC 2014 (*source* https://www.facebook.com/noivadocordeirocomunidaderural/photos/pb.190044884409106.-2207520000.1436819409./721376007942655/?type=3&theater with permission)

During one of the interviews, Canela, Noiva's inhabitant, remembers that the person who installed the Internet told him

> Now you have the world here.
> Initially, he was not impressed and his reaction was as follows:

> I thought: what?!? But at this time the Internet was very slow, and people did not use it as much.

> In 2011, the Internet was installed permanently, from GESAC (the Federal government) which ensured a stable, Internet infrastructure and reasonable speed of connection.

In 2013, a similar pioneering initiative resulted in the installation of a mobile phone repeater mast that enabled the community to have mobile phone access. The community funded this entirely through local contributions, which was made possible due to the collective structure where all inhabitants contribute equally to village projects. The provision of mobile phone access has had a significant impact on the community. Prior to this, they had only one landline pay phone that was often not working. In a field visit, a resident reported:

> It is working very well, and now everyone has a cellphone. Before we had the cellphone mast people would call on the phone, but often it did not work and we couldn't connect to the caller.

> The provision of Internet access, a telecentre and a mobile phone repeater mast, all of which were fairly unusual for a village of the size and location of Noiva, enabled a range of other activities, which impacted on economic and social inclusion of the community.

4.2 Knowledge Networks

One of the key impacts of the introduction of mobile phone infrastructure was the new capacity of local people to use mobile phone communication to support knowledge exchange and development within the village. Currently, almost everyone in the village has a mobile phone, though their primary use is not for voice calls, but for texting. As well as using mobile phones for everyday communication, they are also used to coordinate shared activities. Texts are used to organise collective activities such as the harvest, whereas prior to this, locals had to go knocking from door-to-door to gather people together. For example, Eliene, a local inhabitant, explained that the mobile phone is used 'for everything'. She added that:

> Now, when you have something important and last minute for example, when someone cannot go to D. Matozinha [a village elder], or when you have to organize a task force to harvest anything like that today they connect quickly and everyone knows. Before we had to knock on the door-it was tiring and inefficient.

As part of the community's effort to become economically self-sufficient, they also adopted an entrepreneurial approach and started up a series of micro-businesses. In 1999, the community mobilised its traditional skills in sewing and needlework to start a small-scale lingerie business as a means of generating external income. Initially, the residents travelled to the State Capital, Belo Horizonte, to sell the products door-to-door. But the introduction of the Internet and mobile phone improved access to communication networks that resulted in the possibility of products not only being sold locally but also through the Internet (Zerlotini 2014). In addition, a website was set up[5] to sell goods online to a national customer base. Even though most of the village women have never travelled beyond the state of Minas Gerais, the community also uses the Internet to research global fashion trends. For the business, mobile phone access was also important for sales at local markets and fairs. These require a lot of coordination, and it meant that the women who attend the fair can organise things prior to the fair, and when they attend the fair, they can ring back to the village if needed.

The community is very aware as to the importance of access to information as a way of acting locally but maintaining links to the wider world. This is critical for keeping knowledge within the community. In a conversation, Élida described the importance of the Internet, particularly in terms of keeping the young people from moving away:

> Noiva is "our paradise" but they [the young people] need the technology. They need it to seek information, to discuss new things, to do research. My perception is that they intend to stay here with that communal way of life, but they remain tuned in what happens out of there.

[5]http://www.bhmoda.com.br/index.asp?c=289&m=4&hiper=1&cod_pagina=18777&pag=1.

4.3 Appropriate Contexts for ICTs

In the initial process of bringing Internet to the community, local people were asked how they could use the technology for the development of the community. A key issue that emerged was the potential for the farmers. For example, a tractor driver envisioned how it might help calculating the cost of fuel and a farmer saw the potential in the possibility of knowing the price of produce in the nearest city. As part of a process of appreciating the capacity of the technology to support their livelihood, villagers realised the importance of technology for the management and organisation of the food production in the community. Therefore, farming became one of the key activities that the community identified where they could use ICTs to act more productively. Following the installation of the CIDEC, local inhabitants used the newly introduced technology to collectively plan, manage and distribute the supplies from the community's farming activities.

A further way in which the community has developed quite specific ways of mobilising is through the ways in which they gather together for decision-making and socialising. Community events are generally not one-off occasions, but a series of regularly organised meetings and festivals that take place at a range of different times and locations. Although the primary focus is on face-to-face contact, ICTs are used during the organisation of each of the events. For example, every Friday the inhabitants of Noiva gather together for the 'Night of the Viola'. This new weekly 'tradition' was introduced as an opportunity for people to gather together socially but also as a forum in which to discuss important issues in the community. The event provides an open platform for discussions about issues people feel are relevant to the community; topics range from everyday matters such as farming to more contentious issues such as homosexuality and abortion. Such events (occasions) provide a valuable platform for knowledge sharing and development.

5 Discussion: Power, Knowledge and Appropriate Occasion

In this section, we discuss the correlation between ICT appropriation and everyday practices using as a basis the conditions underlined by Feenberg for technical agency: power, knowledge and appropriate occasion. In doing so, we aim to demonstrate the appropriateness of such conditions within the context in which we are working through a discussion of each of his conditions.

Over the last decade, the introduction of the telecentre (CIDEC), the ICT training within the community and the installation of the mobile phone antennae have contributed to a significant change in both the everyday practices and the overall organisation of the community. From a point of almost complete social and economic isolation, Noiva has acquired a range of ICTs to improve agricultural practices and invest in cultural activities. Their village's unique societal structure has enabled them to broaden their connections with the wider world both online and offline. In many ways, they can be considered to have empowered themselves, and ICTs have been one of the ways in which they have mobilised a 'capacity to act' or technical agency according to Feenberg's definition.[6]

5.1 Power: Mobilising ICT Infrastructure

Up until about 2005, even basic ICT availability was very limited; public payphones were often faulty, individual phone lines were prohibitively expensive, and Internet access was non-existent. Mobile phones were also not available due to both the inadequate infrastructure and the cost of the service. This situation reinforced the marginalised position of the community, as the village was not accessible through either the transport or information infrastructure. Collective community action was critical to securing access to ICTs. Firstly, in the partnership with Vale Foundation, CDI and the government which provided computers, infrastructure and training later on, Internet access, and secondly, in the collective funding of the mobile phone antennae. The community not only actively engaged in bringing the infrastructures to the village, but did so in a way that is appropriate to their own needs. In Feenberg's thinking, power instead equates with being empowered, and so through such appropriation they empowered their community. The important aspect of how they brought in the infrastructure was that they did so within existing networks of power, those of the telecoms providers and local governance. The manner in which the community operates as a collective enabled them to mobilise coherently behind such decisions and for instance collectively raise the money to pay for the mobile phone repeater antennae.

5.2 Knowledge: Mobilising Knowledge Networks

Noiva has sought to diversify and apply formal knowledge to reinforce their own experience, 'the knowledge from below'. Community members have used Internet access to diversify the economic activities that link them to the outside world, and also provide additional sources of income. For example, the products of the lingerie business, run by the community, are sold locally and on the Internet (Zerlotini 2014). Through the telecentre, in particular, the community can access the Internet as a knowledge-building tool, as well as use it to develop new tools

[6]There is one aspect of the way the community acts that we do not deal with directly; that of gender. With the men of the village absent for much of the week, the community is organised and run primarily by the female inhabitants. Although we consider this to be a significant factor in how the community has developed a capacity to act, we do not deal with this directly here.

to support community social events and networks. This effectively combines two sorts of knowledge as outlined by Feenberg: that related to technical issues which he defines as a 'specialised knowledge' and that which comes from experience, which he calls a 'knowledge from below'. A further example is how the community re-imagined its farming activities through ICTs, where both computer skills and mobile phone communication are used to open up capacity, particularly at key times such as during the harvest.

5.3 Appropriate Occasion

The community has created a range of platforms through which they can mobilise ICTs for the benefit of the community. This includes the telecentre, which is used to manage farming activities as well as the community social gatherings that provide a framework for collective knowledge sharing and decision-making. These 'occasions' provide the circumstances in which the application of knowledge and power makes sense (Feenberg 2011), which according to Feenberg is critical to enacting social transformation. Appropriate occasion, in this sense, means not only the point at which technical agency is enabled, but also the circumstances that allow for community members to appropriate ICTs for their own benefit. A final condition, and one which is less salient, is how agency is enabled through the collective use of space in the community. Community spaces have specialist functions, such as a kitchen, a meeting space and a telecentre, but are each collectively owned, run and managed. In fact, the combination of multiple activities located within a series of central spaces has meant that social ties are constantly reinforced through contact. The community places operate as part of a dense network of social and spatial resources. Community spaces such as the telecentre are centrally located spatially, and ICTs are successfully embedded within these spaces. ICTs are not seen as technological devices but are accessed based on how they support and enhance community activities and needs. The shared ownership of the space is also reflected in the shared responsibility for ICT infrastructures and devices, such as the collective procurement of the mobile phone antennae. For Noiva, the sense of the collective, at a spatial, social and economic level, provides the circumstances through which appropriate occasions are enabled through ICT use.

6 Summary

The pervasiveness of ICTs in urban settings has meant that much of the current discussion focuses on cities and cutting-edge technologies, with the consequence that the role of ICTs in rural and marginalised settings is often overlooked or misunderstood. Yet, these contexts have the potential to benefit from ICT use and to overcome conditions of economic, social and spatial exclusion. We discussed

Feenberg's concept of technical agency in terms of how a community can use ICTs to mobilise change within their everyday lives. We introduced Feenberg's three conditions for technical agency in order to systematise an analysis of ICTs: knowledge, power and appropriate occasion. We discussed a case study of a small rural community in Brazil and investigated the role of ICTs in community development through the lens of technical agency. We outlined how Noiva do Cordeiro uses ICT infrastructure to communicate and share knowledge more effectively and also to support activities appropriate to their own context and activities. In this way, they have supported their capacity to act (technical agency), and access to the Internet and mobile phones has been used to counter their geographic and social marginalisation. This has been achieved in a range of ways: through collective empowerment such as mobilising to collectively install ICT infrastructure; through their active use of ICTs within spaces such as the telecentre for knowledge sharing both locally and globally online; and through creating appropriate occasions for how they mobilise knowledge within these power structures.

Acknowledgments The empirical work in the village of Noiva do Cordeiro was undertaken during 2012 and 2013 as part of the work of the research groups MOM-UFMG and Lagear-UFMG and the research Museu Tempo at Universidade Federal de Minas Gerais, Brazil, funded by FAPEMIG, FINEP, CAPES and CNPq. Lorena Melgaço's Ph.D. research is funded by CAPES/REUNI scholarship. The research is part of the Digital Neighbourhoods research project at Plymouth University, UK, funded by an EU Marie Cure grant. We would like to acknowledge the contribution of the researchers Ana Paula Baltazar, Silke Kapp and Viviane Zerlotini from Universidade Federal de Minas Gerais. We would also like to thank the residents of Noiva do Cordeiro, for their participation in the interviews and data gathering.

References

Amin, A. (2002). Spatialities of globalisation. *Environment and planning A, 34*, 385–400.

Bakardjieva, M. (2009). Subactivism: Lifeworld and politics in the age of the internet. *The Information Society, 25*, 91–104.

Baltazar, A. P. (2009). *Cyberarchitecture: The virtualisation of architecture beyond representation towards interactivity.* Unpublished PhD thesis, The Bartlett School of Architecture, University College London, London.

Baltazar, A. P. (2013). *Relato da visita à Noiva do Cordeiro (internal report).* Belo Horizonte: Universidade Federal de Minas Gerais.

Bowater, D. (2014). Brazil's valley of beauties appeals for single men. *The Daily Telegraph,* August 27, 2014. http://www.telegraph.co.uk/news/worldnews/southamerica/brazil/11060282/Brazils-valley-of-beauties-appeals-for-single-men.html

Castells, M. (2007). Communication, power and counter-power in the network society. *International Journal of Communication, 29*, 238–266.

Castells, M. (2011). *The rise of the network society: The information age: Economy, society, and culture* (Vol. 1). London: Wiley.

Cerenza, F. (2009). The women of the village. *Marie Claire*, November 10, 2009. http://revistamarieclaire.globo.com/Revista/Common/0,,EMI102103-17737,00-A+VILA+DAS+MULHERES.html. Accessed January 01, 2015.

de Sola Pool, I. (1990). *Technologies without boundaries: On telecommunications in a global age*. Harvard: Harvard University Press.
Demo, P. (1994). *Pobreza política*. Campinas: Autores Associados.
Demo, P. (2007). Marginalização digital: digital divide. *Boletim Técnico do SENAC 33*, 5–19. http://www.senac.br/informativo/bts/332/artigo-1.pdf. Accessed January 01, 2015.
Feenberg, A. (1999). *Questioning technology*. London: Routledge.
Feenberg, A. (2009). Democratic rationalization: Technology, power and freedom. In David Kaplan (Ed.), *Readings in the philosophy of technology* (pp. 139–155). Plymouth: Rowman and Littlefield Publishers.
Feenberg, A. (2011). *Agency and citizenship in a technological society*. Lecture presented to the Course on Digital Citizenship, IT University of Copenhagen. https://www.sfu.ca/~andrewf/copen5-1.pdf. Accessed September 17, 2014.
Fernandes, R. (2014). A comunidade que nasceu para o mundo. *Jornal Estado de Minas Blog*. http://sites.em.com.br/app/noticias/minas-viva/2013/07/04/noticia_minas_viva,6/a-comunidade-que-nasceu-para-o-mundo.shtml. Accessed January 10, 2015.
Flusser, V. (n.d). *No além das máquinas*. Unpublished paper, Courtesy: Flusser Archiv, Berlin.
Graham, S. (Ed.). (2004). *The cybercities reader*. New York: Routledge.
Graham, S., & Marvin, Simon. (2001). *Splintering urbanism: Networked infrastructures, technological mobilities and the urban condition*. New York: Routledge.
Gurstein, M. (2003). Effective use: A community informatics strategy beyond the digital divide. *First Monday, 8*(12).
Gurstein, M. (2007). *What is community informatics: (and why Does it Matter)?*(Vol. 2). Milan: Polimetrica.
IFAD. (2003). *Fighting rural poverty. The role of information and communications technologies*. http://ifad.org/events/wsis/phase1/synthesis/synthesis_e.pdf. Accessed December 15, 2014.
IT for change. (2014). *Round table on inclusion at the network society*. http://itforchange.net/sites/default/files/Round%20Table%20on%20Inclusion%20in%20the%20Network%20Society-Full%20Report.pdf. Accessed January 01, 2015.
Kapp, S. (2013). *Visita a Noiva do Cordeiro: relato imediato*. Internal Report. Belo Horizonte: Universidade Federal de Minas Gerais.
Latham, R., & Sassen, S. (2005). Digital formations: Constructing an object of study. In R. Latham & S. Sassen (Eds.), *Digital formations: IT and new architecture in the global realm* (pp. 1–34). New York, London, Tokyo, Princenton: Princeton University Press.
Massey, D. (2004). Geographies of responsibility. *Geografiska Annaler: Series B, Human Geography, 86*, 5–18.
Mbarathi, N., & Diga, K. (2014). *Discontinuities in m-banking and development policy: A case study of women in Kenya's agricultural sector.* http://itforchange.net/inclusionroundtable2014/file/view/361/discontinuities-in-m-banking-and-development-policy-a-case-study-of-women-in-kenyas-agricultural-sector. Accessed January 07, 2015.
Melgaço, L. (2013a). *Relato de visita a Santo Antonio do Salto em 17e 18/08/2013*. Internal Report. Belo Horizonte: Universidade Federal de Minas Gerais.
Melgaço, L. (2013b). *Relato de visita a Noiva do Cordeiro em setembro de 2013*. Internal Report. Belo Horizonte: Universidade Federal de Minas Gerais.
Melgaço, L., & Baltazar, A. P. (2011). The museum of today: Toward bridging the socio-cultural divide in Brazil. In P. Abend, T. Haupts, & C. Müller (Eds.), *Medialität der Nähe: Situationen, Praktiken, Diskurse* (pp. 195–208). Berlin: Transkript Verlag.
Mossberger, K., Tolbert, C. J., & Stansbury, M. (2003). *Virtual inequality: Beyond the digital divide*. Washington, DC: Georgetown University Press.
Noiva's Inhabitant. (2013). Interview on 17 September 2013. Not published.
Roper, M. (2014). Town in Brazil made up entirely of women has made an appeal for bachelors (but only those willing to live by female rules!). *The Daily Mail*, August 27, 2014. http://www.dailymail.co.uk/news/article-2735435/Town-Brazil-entirely-women-appeal-bachelors-willing-live-female-rules.html. Accessed January 10, 2015.

Sassen, S. (1991). *The global city: New York, London, Tokyo*. Princeton: Princeton University Press.
Sassen, S. (2006). Sociodigital formations: Constructing an object of study. In B. Schneider & S.-K. McDonald (Eds.), *Scale-up in education: Ideas in principle* (pp. 203–216). Lanham: Rowman & Littlefield Publishers.
Souza, J. (2009). *A ralé brasileira*. UFMG: Quem é e como vive. Belo Horizonte.
Swyngedouw, E. (2004). Globalisation or 'glocalisation'? Networks, territories and rescaling. *Cambridge review of international affairs, 17*(1), 25–48.
Thackara, J. (2001). The design challenge of pervasive computing. *Interactions, 8*, 46–52.
UNESCO (2010). EFA global monitoring report 2010: *Reaching the marginalized*. Paris: Unesco Publishing/Oxford University Press.
Unwin, T. (2009). *ICT4D: Information and communication technology for development*. Cambridge: Cambridge University Press.
Willoughby, K. (1990). *Technology choice: A critique of the appropriate technology movement*. New York: Westview Press.
World Bank. (2001). *Committee for democratization of information technology*. http://info.worldbank.org/etools/docs/library/91628/telecentres/telecentres/workshop/sbt-pdf/case-studies/CDICaseStudy.pdf. Accessed January 10, 2015.
Zerlotini, V. (2014). *Espaços coletivos de trabalho: entre a produção e a reprodução*. Ph.D. Thesis. Belo Horizonte: Universidade Federal de Minas Gerais. http://www.bibliotecadigital.ufmg.br/dspace/handle/1843/BUOS-9LEM5B. Accessed November 12, 2014.

Police, Protests, and Policy in Rio de Janeiro—Mega-Events, Networked Culture, and the Right to the City

Lea Rekow

Abstract This essay looks at the role networked media plays in orchestrating direct action, exposing social injustices, and representing political dissent in relation to the mega-events hosted in Rio de Janeiro in 2014–2016. It examines how independent media and citizen reportage are shaping practices of social organizing and civic mobilization to change the focus of public debate around these events and issues relating to them. It also explores how the government, in turn, is widening its use of digital military applications to monitor citizens and restrict the right to the city under the rhetoric of national security, and how this is impacting on individual freedom of expression, the structure of public protest, and threatening the democratic capacity for social and political impact.

Keywords Protest · Policy · Events · Networked culture · Civic engagement · World cup · Brazil

1 Introduction

Whether part of a formal strategy by the State, or more informally at the individual or collective level, advancing technologies are increasingly integrating into the social fabric of Rio de Janeiro to impact the societal structure of the city. Though face-to-face interaction remains a key foundation for building social movements in Brazil, social and technical networks are amalgamating to form

L. Rekow (✉)
Green My Favela, Rio de Janeiro, Brazil
e-mail: learekow@gmail.com
URL: www.greenmyfavela.org; www.learekow.com

M. Foth et al. (eds.), *Citizen's Right to the Digital City*,
DOI 10.1007/978-981-287-919-6_7

hybrid organizational models that shape and facilitate collective action and State response. The use of technology is dramatically transforming how and where security forces are deployed, how they surveil, and how they are being surveilled. In Brazil, this specifically coincides with a major shift in access to information communication technologies (ICTs), which are now comprehensively employed by the various branches of government and military police to monitor and target citizens and quash protest activities. ICTs are also used by citizens, activists, and journalists to promote and coordinate civil actions, record events, and use as evidence in cases of police misconduct.

The protests that erupted in Brazil in 2013 and continue to smolder in Rio de Janeiro and elsewhere focus on a range of grievances that include high levels of police brutality, political corruption, income inequality, a lack of public services, and the excessive cost of hosting mega-events such as the 2014 World Cup and 2016 Olympics. Rio's urban policies surrounding these mega-events have played a key role in civic unrest, especially in regard to issues such as public transport, housing, education, and public security. Networked protest culture has galvanized around these issues to catalyze civic mobilization and expose numerous violations of spatial justice throughout the country.

The scenario in Brazil mirrors that of other protest landscapes globally, in that they are typically coordinated by young networked citizens connected through non-hierarchical frameworks and enabled through ICTs that provide access to higher levels of civic engagement through exposure to increased circulation of diverse information (Gil de Zúñiga et al. 2012), including user-generated content (Östman 2012). The social impact of these activities is aimed at broad political reform (Loader et al. 2014), however in this regard, the outcomes have been limited at best.

Government response has been predictably swift and harsh, impacting on cyber policy, the personal security of protesters, and urban mobility. Brazil is already one of the most dangerous places in the world to be a media professional (Reporters Without Borders 2014). Now, more than ever, digital activists are increasingly at risk and significantly vulnerable to retaliatory mechanisms implemented by the State.

The multiplicity of ICT sources that are crafting both civil engagement and State response on the battlegrounds for public space (both online and offline) is sufficient to ensure neither civic freedom or quash political dissent. Therefore, this essay aims to illustrate how, on both sides of the cultural–political divide, ICT flows connect through political conflict to influence Brazil's public security framework. It catalogues a series of activism and repression events to reveal trends that:

(a) exemplify the sociopolitical role that networked publics play in the quest for institutional reform with the goal of securing the right to the city.
(b) explain the tactics employed by the State to suppress these actions in order to protect its economic development interests and partnerships.

Its purpose is to impart an understanding of the cat-and-mouse nature of these event trends and to assess their influence on policies, citizenship, and Brazilian democracy. In other words, it aims to describe how the oppositions of cyber politics are currently playing out in public space and how this relates to the economic development and public security agendas surrounding Rio's mega-events.

2 Architecture of a Mega-Event: A Cidade Para Quem?

Rio de Janeiro is a city that can be defined as much by its excess of private wealth and privilege as by its immense concentrations of slums. Its policies promoting rapid urbanization, accelerated by its successful bid to host the World Cup and Olympics, have expanded a landscape of profitable capitalist activity and associated military expenditure.

Considering that Brazil, and Rio in particular, has one of the highest levels of income inequality in the world—with one in five people living in favelas (slums)—the protest grievances regarding mega-events are justifiable. In a country where half of the population lives on incomes of less than $130 per month (IBGE 2010), the decision to spend $11 billion to stage the most expensive World Cup ever (Porta da Copa 2014) deserves to be deconstructed.

Though it is difficult to measure the total cost to Brazil, public works spending around the World Cup and Olympics could reach as high as $1 trillion. Just the expense of the 157,000 security troops deployed during the Cup cost Brazil close to $1 billion. The event only generated $4 billion in total revenue—mostly from the sale of broadcast rights and advertising, with the majority of the profits going to FIFA (now under investigation for corruption in connection with the event). The Olympics is expected to yield about the same, with most profits benefiting the International Olympic Committee. Even if those profits were going back into the country, that still leaves a massive investment shortfall, which begs the question, who reaps the benefits?

Rio's Unidade de Polícia Pacificadora (Police Pacification Unit or UPP), augmented by other Special Ops forces (BOPE and Shock Troops), plays a key role in protecting the city's public/private economic development interests that link to the mega-events. The UPP was specifically established in 2008 to take military control of favela territories in areas located close to international cultural and sporting venues and affluent neighborhoods.[1] To date, 9543 UPP police have been deployed in 38 favelas close to hotels, tourist spots, sporting venues, and the city's wealthy Zona Sul district. These military police forces are employed to combat public protest, enforce territorial control over strategic locations, and safeguard the fast-tracked urbanization projects being implemented in preparation for the mega-events.

[1]The remaining 700+ favelas of Rio have not been affected by pacification. Most remain under the control of the drug trafficking gangs that have governed them for decades.

Several multinational giants of the surveillance sector profited from World Cup contracts to provide everything from security scanners to remote tracking software and data security services. These contracts supplied law enforcement divisions, telecommunications sectors, intelligence communities, and defense agencies with whatever they needed to engage in high-tech electronic investigations, network intelligence gathering, and military applications. Surrounding Rio's Maracanã sports stadium, missiles were placed on apartment roofs, and airspace surveillance and defense equipment was installed on terraces along with teams of Armed Forces and Special Ops troops.

Odebrecht (which won contracts for four of the stadiums) and its subsidiary Mectron (contracted to upgrade the Army's air defense capabilities for the event) emerged as the true victors of the World Cup. In 2013, the conglomerate's gross revenue increased 16 % to $31.4 billion, making it the largest closely held company in Latin America (Schmidt 2014). Odebrecht is one of the 15 richest families in the country—families that collectively control about $122 billion, or approximately 5 % of all the wealth produced in Brazil.[2] It is one of the largest contributors to the country's major political parties and is currently under investigation for political corruption.

The overwhelming scale of corruption in Brazil is huge. In 2010 alone, corruption was estimated to have cost the country $53 billion—2.3 % of the country's GDP. This enormous amount of misappropriated taxpayer money could otherwise have been spent to improve Brazil's underequipped health care system—by adding 327,000 hospital beds (an 89 % increase); its substandard public school system—by providing 16 million spaces for children (a 47 % increase); and its lack of infrastructure—by installing sewerage systems for 23 million slum dwellers (a 103.8 % increase); and to provide housing for three million families (FIESP 2010); or to better serve the 250,000 families that have been forcibly evicted from their homes, mostly from favelas, in order to secure areas near sport stadiums for mega-events.

For the many who are not served by Rio's rapid urbanization, claiming the right to the city—how it is being shaped and for who (Harvey 2013)—has become the fundamental underpinning of the protest movement and its call for collective human rights. Independent media and networked publics have enabled an explosion of reportage, blogging, and debate around the hidden mechanisms and impacts of the mega-events and how they connect to the economic architecture of the State—or "Brazil Inc." as it sometimes referred to. The use of Facebook, Whatsapp, and Youtube has been a key to the coordination and supersizing of protests. The role that networked collectives of citizen journalists have played in making use of ICTs to bypass the mechanisms of Brazil's old mass media monopoly, not only to organize and mobilize protest actions offline, but to record and publicize social injustices such as police misconduct, has been substantial. This is

[2]In Brazil, just a few more than a hundred families control more than 10 % of the GDP, and the top 5000 richest families (0.0001 % of the population) have an accumulated wealth equivalent to 42 % of GDP.

similar to other global protest situations where ICTs have become fully integrated into the communication protocol of political expression (Wheeler and Mintz 2012).

3 The Mediascape of Protest

The public space of the city has always been a space of contention (Goheen 1998). Urban activists see public space as the grounds where social dynamics—participatory and spontaneous politics, popular claim and counterclaim, and the voice of the underrepresented and emergent communities (Amin 2006)—manifest, shape, and are shaped by both material and non-material culture. In relation to Rio de Janeiro's mega-events, public space is increasingly under threat from private enclosure, forced displacement and land grabs, the use of excessive military force, political corruption that ties to private economic interests, and increased civilian surveillance and repression measures (Fig. 1).

In the digital realm, networked publics are using the most basic forms of social media and mobile technology to effectively document, coordinate, circulate, and

Fig. 1 Protest performance draws attention to the link between the 2014 FIFA World Cup and the sexual exploitation of up to 5 million Brazilian children who were pimped out for the price of a packet of cigarettes during the month-long event

expose information about these threats and the policies that enable them. Brazil is the fifth most wired country in the world. Approximately 80 million social media users make it one of the largest users of social networks globally. Though poverty and illiteracy contribute to a massive digital divide in home Internet access (Patry 2014), the country boasts 278.1 million mobile phone subscribers (Teleco 2014), representing a third of all users in Central and South America. The almost ubiquitous access to this basic networked device has brought with it transformative opportunities to expand social interactions and relationships relating to the politics of defining public space (Lim and Kann 2008).

One of the first independent journalists to use mobile technology as an effective tool for citizen reportage in Rio de Janeiro was Rene Silva, a young student who used Twitter to share information about the violent conflict in Rio's Complexo do Alemão during the UPP's aggressive military "pacification" of the favela which began in 2010. Silva, whose followers swelled to more than 50 thousand as military troops and tanks rolled into occupy his community, was one of the few voices that reported from the inside. As the area was physically isolated during the operation, mainstream media also began to use Silva's tweets as a news source. The teenager later created the *Voz das Comunidades* Web site, an online portal where users share knowledge about favelas. *Voz das Comunidades* now has 159,000 Twitter followers and 35,000 Facebook followers. Within the favelas, these types of social media initiatives continue to grow.[3]

The evolution of this sort of frontline citizen journalism came to prominence with the nationwide mass protests of June 2013. The demonstrations, largely coordinated through the Anonymous hacktivist collective's Facebook page, also led to the rise of the guerrilla journalist collective, Mídia NINJA, who used basic networked platforms and mobile technology as a response to, and a protection mechanism against, the police brutality perpetrated on demonstrators. Mídia NINJAs recorded and live-streamed almost every conflict with police in the anti-corruption demonstrations that called for the impeachment of Rio de Janeiro's Governor Sérgio Cabral and Mayor Eduardo Paes. At the same time, the NINJAs were at the forefront of the long-running occupation of City Hall and in the marches at the headquarters of the Globo media conglomerate (Rekow et al. 2014), which rose to power during the dictatorship to become Brazil's dominant media corporation and the second largest income-generating commercial television network in the world. Originally formed in 2011, the Mídia NINJAs grew out of the communications arm of a cluster of student collectives who covered cultural events, incidents in the favelas, small-scale protests, and other political events that the mainstream media ignored. Mídia NINJA now has over 2000 collaborators, more than 300,000 Facebook followers, and 40,000 Twitter followers, and persists in exposing a range of excessive and illegal police actions, infiltrations, and unlawful arrests.

[3]Silva has become internationally recognized for his achievements in building inclusive digital communities. He has participated at a digital communications symposium at Harvard University and a journalism exchange with an NGO in New York. He was even one of the torch bearers at the London Olympics. He still lives in Complexo do Alemão.

For the citizen journalists that continue to take advantage of the publicity generated around the city's mega-events, their commitment to protest comes at a high price. They have been beaten, shot at, tear-gassed, pepper-sprayed, dispersed with water canons, and had stun grenades thrown at them. Together with the tens of thousands of others that have crowded the streets, their efforts to withstand subjugation have been arduous. In a single day in June 2013, attempts by the State to suppress the protests included the pepper spraying of a bus, the tear gassing of several downtown bars and restaurants, and a police attack on a hospital where Shock Troops discharged rubber bullets and tear gas. The Federal University's Law School and the Institute of Philosophy and Social Sciences were surrounded by military police, isolating 700 civilians, including 400 students. Street lighting was shut off in the central business district, metro stations were closed, and the city's IBM high-tech operations center—responsible for coordinating and deploying security, defense, and emergency agencies—deceptively switched its camera feed of a major downtown avenue to depict traffic in an entirely different neighborhood. Information disseminated through independent media channels gave the only indication of the extent of the police violence, which included the deployment of 7500 armed police units utilizing tanks, horses and dogs, and the dispersal of over 2000 tear gas bombs (Rio on Watch 2013).

Just one month later, whole neighborhoods were isolated by military troops during Pope Francis' visit to Rio, another mega-event hosted by the Marvelous City. Lawyers were assaulted, journalists beaten and arrested, civilians shot, and children pepper-sprayed in conflicts that persisted throughout the week-long event. Thousands gathered to demand government impeachments, to protest the $53 million of public funds spent to host the Pope, and to demand attention for the disappearance of citizens by police. Military troops cut street lighting, cordoned off neighborhoods, and used concussion grenades, tear gas, pepper spray, water jets, tasers, and rubber bullets on the crowds. One protestor, Bruno Telles, was beaten, tasered, shot, and arrested for throwing a molotov cocktail that burned two police officers. Photographs and video footage shared through social media networks later revealed that the perpetrator was most likely a plainclothes police officer inciting violence, and the charges were dropped.

During the World Cup, the aggression intensified. While an almost exclusive rich white audience filled Rio's Maracanã stadium at an average cost of $1000 a seat (or $7000 for the final match), outside the venue extreme violence was being perpetrated. On the night of the first match at Maracanã, when Argentina played Bosnia, journalists picked up spent shell casings after police fired live rounds at protestors. A few days later, as the world cheered their teams on during the Chile versus Spain game, the UPP killed 25-year-old Afonso Linhares while he refereed a community football match in Rio's Manguinhos favela, just four kilometers away. The same UPP officers had killed another 19-year-old boy only the month before in the same community. Just days later, more deaths followed in nearby Complexo do Alemão. These incidents were almost exclusively reported on by independent journalists. The fact that youth were being murdered and shot at by police during the Cup was virtually ignored by mainstream media.

The mediascape of independent reportage reveals the violent reality of living in Brazil. State violence is particularly exaggerated in Rio, where police are responsible for approximately 1100 extrajudicial deaths per year (Human Rights Watch 2009). Young black men make up the majority of these fatalities as the victims of "resistance killings" (Alston 2010). Since the beginning of 2014, the numbers have risen dramatically, with resistance killings up by 59 % from the year before. They reached their highest point in June during the World Cup, when police killed 44 people throughout the city (Fórum Brasileiro de Segurança Pública 2014).

The UPP and their Special Ops counterparts are responsible for numerous instances of suspicious deaths inside pacified favelas. As a result of increased publicity surrounding the mega-events, and as a consequence of independent media coverage, many of these deaths and disappearances have gained international attention and ignited public outrage.

Douglas Pereira, a professional dancer who was found beaten to death in a school during a police operation in Rio's Pavão-Pavãozinho favela in April 2014, sparked a massive protest that spilled into the streets of Copacabana. Favela residents set barricades on fire and clashed with riot police. Traffic was disrupted, the metro was shut down, and power was cut to the neighborhood. Another civilian was shot in the head during the conflict. The following week, the police used tear gas and rubber bullets to disperse the crowd after Pereira's funeral. Mídia NINJAs were front and center in both conflicts, providing information to international media outlets and bringing the UPP and the World Cup under scrutiny.

Another prominent civilian death involving the UPP was that of Amarildo de Souza, the 42-year-old bricklayer who disappeared while in custody at police headquarters in the Rocinha favela in July 2013. His disappearance prompted a social media campaign *Onde está Amarildo* (Where is Armarildo) that gained global traction and sparked protests nationwide. Twenty-five police officers were subsequently charged with the bricklayer's torture and the concealment of his corpse. Prosecutors allege that de Souza was killed while being interrogated by the UPP who used tasers and waterboarding on the victim, who was epileptic, before suffocating him with a plastic bag. Out of the 6034 disappeared people recorded in Rio between November 2012 and October 2013 (Carpes 2014), the case of Amarildo de Souza was the only one investigated. This official inquiry came about solely because of the light shone on the incident through independent reportage and the international outcry that it generated.

Yet another example of gross police misconduct brought to light through citizen reportage took place in March 2014 when Claudia Silva Ferreira, a young black woman, was shot on her way to the grocery store by police officers involved in a raid targeting drug traffickers. The police threw Ferreira into the trunk of their hatchback where she lay for two hours before they sped away with her still in the car. The trunk opened when it hit a bump in the highway, catching Ferreira's leg as her body was thrown out of the moving vehicle. She was dragged along the asphalt for almost a thousand feet until the officers stopped and stuffed her back into the trunk. The event was captured on video via cell phone by an individual traveling behind the police vehicle. The footage went viral and subsequently led to the arrests of three police officers, two of whom had been involved in a number of other police killings—Warrant Officer Rodney Archanjo had been named in three homicides involving police intervention, and Lieutenant Adir Machado had been named in a staggering thirteen (Fig. 2).

Citizen watchdogs are also monitoring social media posts by police. On February 2, 2014, a photograph showing police posing next to several dead youth lying in an alleyway inside a favela in Rio's North Zone was uploaded to the BOPE Military Police's Facebook page. The gruesome imagery of the blood-soaked children, surrounded by the officers who had executed them, had been posted alongside the comment, "We don't go into favelas to die. We go into kill." The image was taken down after a range of media outlets criticized the post, though the incident was never investigated.

These notable examples of citizens monitoring and recording police actions illustrate how networked publics may be one of the few means with which to provide evidence and stimulate social justice dialogue of police atrocities, especially those inside the favelas of Rio. They also exemplify the security dynamic between civilian and State actors, and how this dynamic is affected by the use of ICTs in the hands of the public. They reveal that many of the citizens who are using ICTs for political or civic purpose are not connected to any activist or protest movement, and because of this, the lines between reportage, analysis, consumption, and

Fig. 2 Protesting the death of Cláudia Silva Ferreira outside FIFA accommodations at the Copacabana Palace Hotel on opening day of the 2014 World Cup

distribution of information can be less clearly defined. Finally, they demonstrate that however vital ICT usage is to combating social injustice in Brazil, its impact on the sociopolitical landscape is, at best, extremely limited.

4 The State Response: A Guarantee of Law and Order

The growth of networked media and citizen reportage is not, of course, lost on Brazil's political machine. Not only are ICTs playing a critical role in expanding democratic voice, they are also instrumental in repressing political dissent (Ziccardi 2013).

Rio's UPP and other military police forces are considerably aided by investments in technology, evidenced and most visible in Rio's "IBM Intelligent Operations Center." The Center coordinates a citywide data network that gathers and analyzes hundreds of video feeds, environmental monitoring, traffic flows, and emergency services. It also directs federal and state police, deploying shock troops to actual, perceived, or potential crises with bomb-busting robots, surveillance drones, anti-aircraft tanks, helicopters, and dog teams. Merging with the IBM Operations Center, Rio's Integrated Command and Control Center—built at a cost of more than $50 million to the State (Portal da Copa 2013)—was designed to monitor public movement, coordinate security forces, and provide technological and telecommunications support for police.

The architecture of big data for mass telecommunications interception was launched in 2011. Though it is expected to take ten years to fully implement, the project has been accelerated because of the mega-events. It is estimated to cost $3.1 billion (Fonseca et al. 2013). The project involves networking radar, communications systems, unmanned aerial vehicles, and networks of data sensors, all synchronized through the Command and Control Centers. These centers also coordinate with the Brazilian Intelligence Agency, which has dramatically increased its social network monitoring as part of a strategy intended to scrutinize protest activities.

The authorities use software similar to that employed by the North American National Security Agency to troll and retrieve civilian social media communications. Information gathered by the Brazilian Army's Center for Cyber Defense is being used to monitor communications related to protest activities and to locate activists. The information is relayed to the Federal Police, State Departments of Public Security, and the Federal Public Prosecutor's Office. Supported through federal legislature, these procedures are legally justified under the pretext of national security.

Since 2013, Brazilian Congress has tabled 16 bills aimed at defining cyber and civil actions as terrorism. Whether or not any eventually pass, the Ministry of Defense has already issued a "Guarantee of Law and Order" that provides the armed forces with an "Operational Plan" for "Disorder control in the urban environment" and refers specifically to "social movements, opposition or protest [that] undermine public order." It also seeks to "define the flow of information to the general public, especially to the media" and within a cyber defense structure, to "Coordinate operational systems that interact with the dissemination of propaganda and counter-propaganda and public information" (Department of Defense 2013).

During the World Cup, the federal government also created a "Crisis Cabinet" to control demonstrations. The Cabinet, coordinated by the Attorney General's Office, together with the Federal Public Ministry, the Ministry of Justice, and the Military Prosecutor, was formed so that swift action could be taken if protests arose. The Cabinet was, in part, established in response to a demonstration in Brasilia in May 2014, when a police officer was hit with an arrow shot by an indigenous protestor near the National Stadium. Photographs of the event went viral, and fears arose that the incident would put forward an image of Brazil as being unsafe (Portal Brasil 2014).

Brazil's security agencies are, in part, enabled by a landmark Internet civil rights law (Marco Civil da Internet 2014), signed into effect by President Rousseff in April 2014. The legislation has been applauded for safeguarding user and service provider rights—including freedom of expression, interoperability, open standards, personal data protection and accessibility, multi-stakeholder open government data, copyright, and general net neutrality. However, the law also includes a data retention obligation that compels Brazilian Internet communications and content service providers to retain specific metadata for a period of up to 12 months. The data are accessible pursuant to court order. This mandatory data retention brings risks to civil liberties with it. Social media has become a tool for the government to use to both monitor and manipulate information, especially that which is connected to protest activities, under the guise of preserving law and order.

Brazil's intelligence agencies are responsible for gathering an enormous amount of online information that has been used to justify a series of protester arrests. The day before the opening of the Cup, police raided activists' homes and arrested eight people in Rio de Janeiro. The day preceding the final match, police issued warrants for an additional 60 activists, arresting 23 by the day's end. The operation was conducted by the city's Police Repression of Computer Crimes division. Eighty police officers from 25 stations, augmented with aerial support, arrested and imprisoned protesters for possessing gas masks and flags. The arrests included that of Camila Jourdan, a philosophy professor at the federal university in Rio; Elisa Pinto Sanzi, a leader of the collective Independent Popular Front; and members of the Internationalist Front for the Homeless; Anonymous Rio; Black Bloc; and the Popular Revolutionary Student Movement. Also arrested was 45-year-old human rights lawyer Eloísa Samy, who had been providing legal representation to protestors. Samy's home was raided, and her computer and phone were seized. She has since gone into hiding. Those that were unable to be located are now considered escaped prisoners. Sixteen others, including three adolescents, one lawyer, and two mothers, were taken at gunpoint, without warrant or charge, to headquarters in the trunks of police cars (RioNaRua 2010).

The 23 protestors were charged with "forming an armed criminal gang" under Article 288 of the Penal Code, a law which prohibits "associating with three or more people for the specific purpose of committing crimes" (Law No. 12,720, Article 288-A 2012). If convicted, the law mandates a one- to three-year prison term. The arrests were based on information obtained from telephone, informatics, and telematics records. The Institute of Human Rights Defenders claim the methods employed for gathering the information breached confidentiality laws (RioNaRua 2014). According to the lawyers, the warrants were issued in order to prevent *potential* crimes—on the *suspicion* that individuals may eventually commit a crime at some point in the future based on the fact that they had participated in protests in the past. The 2000 pages of data produced by the police that laid the foundation for the warrants included the name of Mikhail Bakunin, a Russian philosopher and founder of collectivist anarchism, who died in 1876. The "suspect" had been quoted by a protester in a phone conversation intercepted by police (Revolution News 2014).

There have been many questionable arrests made in connection to the protests in Rio. One of the first was that of 25-year-old Rafael Braga Vieira, a young homeless man who was imprisoned in June 2013 for carrying cleaning liquid. Vieira was poor and black and not connected to any protest movement, but was picked up and charged because he had a bottle of Pine Sol, which is considered an "explosive or incendiary device" (Law No. 10,826/03). He was sentenced to five years in Rio's notorious Bangu prison, where he is incarcerated in an overcrowded cell with more than seventy others, and is entitled to receive two hours of sunlight a day. To date, he is the only person convicted for crimes related to the protests.

In a separate, yet equally bizarre incident, Caio de Souza and Fábio Barbosa were arrested in relation to an accident involving a bottle rocket that killed news cameraman Santiago Andrade in a protest in Rio in February 2014. The cameraman was struck in the head by the firework allegedly lit by a protester. Barbosa surrendered after being identified by images provided by TV Brasil—the first time any alleged perpetrator of any protest crime had ever voluntarily turned themselves in. Barbosa entered the police station with private defense attorney Jonas Tadau Nunes. Nunes was also attorney for Natalino and Jeroniminho Guimarães, both local politicians and militiamen believed to be responsible for the massacres of several people. The notorious Guimarães brothers are linked to the same political party as the Governor and the Mayor of Rio, both of whom have been targets of anti-corruption protest efforts. Nunes proceeded to implicate human rights advocate Marcelo Freixo in the incident by releasing a statement to O Globo media, who ran the story unsubstantiated. Freixo was a strong running opponent to the Mayor, and the person responsible for the arrest of the Guimarães brothers and two hundred other militiamen belonging to the same political party (also defended by Nunes). O Globo then provided a photograph of an additional alleged perpetrator who Nunes claimed he could somehow identify. Breaking official protocol, he accompanied police on a sting operation, along with O Globo journalists, to arrest Caio de Souza. Nunes then became de Souza's defense attorney. In a clear conflict of interest, Nunes is now defending both Barbosa against de Souza and de

Souza against Barbosa. Both await trial on remand. If convicted, the two face up to 35 years in prison.

Even though there had been at least 118 cases of assault on journalists in protests before this incident (Pasolini 2014), mostly at the hands of police, O Globo media singled out the cameraman's accident as a way to demonize the protest movement.

Historically, the Globo network's news content has exerted tremendous influence within the political sphere. Though media bias is not uncommon, it is rare for demonstrations to specifically target media organizations in the way Brazilians did when they attacked and vandalized the headquarters of Globo TV in Rio during the 2013 protests. Globo owners, the Marinho brothers, are Brazil's richest family with a combined worth of $26 billion. Growing out of the era of the dictatorship, the conglomerate has been accused of aligning themselves with political power as a way to protect their interests. In a country long dominated by such a mainstream heavyweight, independent reportage provides a crucial counter voice. And though online viewership through Indymedia channels remains considerably less than any of Brazil's major networks, groups such as the Mídia NINJAs have substantially contributed to changing the face of Brazil's mediascape as a whole.

Despite activists continuing to build cultural capital through the use of independent media, a range of limitations including fragmented access to the Internet, disparate sociopolitical concerns, and the very real threat to individual safety and freedom have resulted in the protest movement being unable to attract a wider base. Though protests were not able to leverage broader support, some policy changes did occur as a result of demonstrations. They included the revocation of Bill PEC-37, which would have limited the capacity to investigate officials, and the pushing through of Law No. 12.858/2013, which reallocated petroleum royalties to fund health and education.

Though social media has impacted on how the public is mobilized and how that plays out in the public sphere, increasing threats to the protest movement and its actors continue to emerge. Even journalists not connected to protests are at risk of punitive government recrimination. Rebecca Mafra was one of five journalists who had arrest warrants issued on July 11, 2014 (the eve of the World Cup), for defamation of a politician on a social network (Revolution News 2014). Mafra's home was raided and ransacked by military police. Her electronic equipment, including camera and memory card, was confiscated for allegedly posting comments against presidential candidate Aécio Neves on a social media site. Mafra denies the charge, claiming she has never posted comments about Neves, that she has no party affiliation and is not a militant, and that her reportage focuses on cinema. Nevertheless, Mafra is being investigated by the Public Prosecutor's Office for cyber crimes at the request of Neves, for allegedly placing irrelevant reader comments on social media news sites. The Prosecutor's Office claims it will offer evidence that the comments were posted in order to manipulate results in search engines like Google.

Threats of personal recrimination make the unforeseen stakes of using social media high. The State's random interpretation of information retrieved from social networks presents a complicated relationship in regard to the sharing of digital information. The ability of the Brazilian political machine to use military applications to remotely or physically access personal devices and the information on them, or to disable media and communications for whole regions, poses a radical threat to democracy, one that is defended at a national level and shrouded under the rhetoric of anti-terrorism.

5 Conclusion

Clearly, the technologies that empower the State to monitor social media activity are increasingly effective at neutralizing both cyber activism and, by extension, democratic voice. State-level cyber security operations, such as those described in this essay, rely not only on the technical operations conducted in cyberspace, but on the policies, strategies, military practices, national security funding, and private industry investments that surround and support them (Lemieux 2015). Attempts to engineer social change through networked political engagement have produced violent repercussions for civilian actors and cannot be considered a safe political means with which to protect the public sphere or bring about institutional reform. Brazil is by no means unique in this regard, but is merely one example of a global trend whereby government actions aimed at limiting digital liberties are contributing to the erosion of public spaces as sites for effective democratic engagement. This erosion of public space is attributable to the multifaceted interconnections between the State and their private development interests and agendas; the excessive appropriation of public resources toward both national security funding and the military-grade cyber-monitoring technologies it employs; and a judicial system that is compliant in all of the above.

The democratic paradigm assumes that one's voice will be represented by one's elected government, and therefore, the interests of the population will be proportionally represented. But in democracies with large disparities between wealth and poverty that serve a small but powerful constituency, public voice can be more easily obfuscated, hidden in plain sight, or simply ignored. In the digital realm, participating in the social mediascape, whether political or not, means making one's own contribution to a mass surveillance system that is hardwired into corporate and government infrastructure. It also means making oneself vulnerable to an unstable and dangerous world in which data can be skewed, lost, accessed, invented, or used as a weapon to protect State interests and create paranoia through a militarized bureaucracy.

Free speech, a free press, and the right to peaceful protest, including the right to form and join associations or groups, are foundational to democracy in order to inform political debate and ensure public accountability and transparency of government. Article 19 of the Universal Declaration of Human Rights theoretically protects the political right "to freedom of opinion and expression," the right "to hold opinions without interference," and the right "to seek, receive and impart information and ideas through any media and regardless of frontiers." Whether in Brazil or elsewhere, legislation is increasingly eroding our basic democratic freedoms by criminalizing these rights through the theater and rhetoric of national security. More and more legal transgressions continue to overstep the mark in regard to digital communications, bringing great harm to individual lives.

It is not only weapons that militarize the police. Inside the world of digital communications, war has been embedded as a state of mind that is fixing itself into policy to permeate and threaten civil liberties. This is an increasingly global phenomenon—a way of dealing with social problems and perceived social problems through militarized force that is supported with the tools of mass surveillance and justified under the pretext of preserving law and order. The social value of this kind of indiscriminate policing is ambiguous at best. What is clear is that this is a mechanism of control employed by a Police State.

It is no surprise that all are not all treated equally under the law. Activists and the poor are institutional targets in the fight for freedom of expression, and State protection is not designed to result in peace or justice, but rather to protect itself by harassing and imprisoning its citizens under a facade of defending individual rights. Now, one of the greatest challenges in preserving the right to the city is safeguarding cyber participation and, by extension, freedom of expression and material participation in the public sphere everywhere.

References

19th Century Anarchist Bakunin, Investigated By Brazil's Police As "Suspect", Revolution News. (2014, 29 July). http://revolution-news.com/19th-century-anarchist-bakunin-investigated-by-brazils-police-as-suspect/.

28 ativistas presos às vésperas da final da Copa, RioNaRua. (2014, 12 July). http://rionarua.org/reportagens/28-ativistas-presos-as-vesperas-da-final-da-copa/.

Alston, P. (2010). *Report of the special rapporteur on extrajudicial, summary or arbitrary executions, follow-up to country recommendations—Brazil, United Nations General Assembly* (pp. 5–6). http://www2.ohchr.org/english/bodies/hrcouncil/docs/14session/A.HRC.14.24.Add4.pdf.

Amin, A. (2006). *Inclusive cities: Challenges of urban diversity*. Woodrow Wilson International the Center for Scholars, the Development Bank of Southern Africa and the CCCB.

Amnesty International Brasil (Direito à moradia adequada). (2011). http://www.amnesty.org/en/node/25117.

Arbitrary detentions on the eve of World Cup final expose serious rights violations in Brazil, Mídia NINJA. (2014, 12 July). https://ninja.oximity.com/article/Arbitrary-detentions-on-the-eve-of-Wor-1.

Bittencourt, G. (2013). *CICC do Rio de Janeiro: Tecnologia para segurança pública, Guia das Cidades Digitais, 10 July*. http://www.guiadascidadesdigitais.com.br/site/pagina/cicc-do-rio-de-janeiro-tecnologia-para-segurana-pblica.

Brazilian Discontent Ahead of World Cup: President Rousseff Gets Poor Marks on Key Issues, Survey, Pew Research Centre. (2014, 3 June). http://www.pewglobal.org/2014/06/03/brazilian-discontent-ahead-of-world-cup/.

Brazil Federal Government Statistics for FIFA World Cup, Portal da Copa. (2014). http://www.copa2014.gov.br/pt-br/brasilecopa/sobreacopa/matriz-responsabilidades.

Brynskov, M., Bermúdez, J. C., Fernández, M., Korsgaard, H., Mulder, I., Piskorek, K., Rekow, L., de Waal, M. (2014). *Urban interaction design: Toward city making* (pp. 43–46). Amsterdam: Floss Manuals.

Carpes, G. (2014). Desaparecidos e esquecidos, Agência de reportage e jornalismo investigativo, 26 February.

Central Intelligence Agency (CIA) GINI Index. (2013).

Decree of Law No. 2848, Article 288, paragraph 1, 7 December 1940. Article 288 paragraph 1 of the Decree of Law No. 2848 (1940) was amended by Law No. 12,850 (2013). Article 288-A was amended by Law No. 12,720 (2012) to include "forming, organizing, integrating, maintain funding, or organizing paramilitary, private militia, group, or squadron for the purpose of committing any of the crimes provided for in the Penal Code." http://www.jusbrasil.com.br/topicos/10602025/paragrafo-1-artigo-288-do-decreto-lei-n-2848-de-07-de-dezembro-de-1940.

Federação das Indústrias do Estado de São Paulo (FIESP). (2010, 13 May). http://www.fiesp.com.br/noticias/custo-da-corrupcao-no-brasil-chega-a-r-69-bi-por-ano/.

Fonseca B., Mota J., Bodenmüller L., & Viana N. (2013). Brazil becomes hot market for surveillance technology ahead of World Cup, 27 September.

Fórum Brasileiro de Segurança Pública, 8º Anuário Brasileiro de Segurança Pública. (2014). Creative commons BY-NC-SA.

Gil de Zúñiga, H., Jung, N., & Valenzuela, S. (2012). Social media user for new and individuals' social capital, civic engagement, and political participation. *Journal of Computer-Mediated Communication, 17*, 319–336.

Goheen, P. G. (1998). Public space and the geography of the modern city. *Progress in Human Geography, 22*(4), 479–496.

Guarantee of Law and Order, Department of Defense, Brazil. (2013). Chapter I, section 1.2 cites Article 142 of the Federal Constitution (1988) that was adopted at a domestic infrastructure level with the enactment of Law Supplement No. 97/99. See in particular Annex G, Chapter 3.1, Chapter 3.6. http://www.defesa.gov.br/arquivos/File/doutrinamilitar/listadepublicacoesEMD/md33_m_10_glo_1_ed2013.pdf.

Governo do Rio de Janeiro Unidade de Polícia Pacificadora (UPP) website. http://www.upprj.com/.

Harvey, D. (2013). *Rebel cities: From the right to the city to the urban revolution* (pp. 3–7). London: Verso Press.

Instituto de Segurança Pública, Rio de Janeiro. (2014, September). http://www.isp.rj.gov.br/Conteudo.asp?ident=150.

International Institute for Democracy and Electorial Assistance. (2014). http://www.idea.int/vt/countryview.cfm?CountryCode=BR. Law 10,826/03, Article 16, paragraph 3. http://presrepublica.jusbrasil.com.br/legislacao/98027/estatuto-do-desarmamento-lei-10826-03.

Lemieux, F. (2015). *Current and emerging trends in cyber operations: Policy, strategy and practice*. USA: Palgrave Macmillan.

Lethal Force, Police Violence and Public Security in Rio de Janeiro and São Paulo, Human Rights Watch report. (2009). New York, p. 31. http://www.hrw.org/sites/default/files/reports/brazil1209web.pdf.

Lim, M., & Kann, M. E. (2008). Politics: Deliberation, mobilization, and networked practices of agitation. In K. Varnelis (Ed.), *Networked publics* (pp. 77–107). Cambridge: MIT Press.

Loader, B. D., Vromen, A., Xenos, M. (2014) *The networked young citizen: Social media, political participation and civic engagement (Routledge Studies in Global Information, Politics and Society)* (pp. 1–15). New York: Routledge Press.

Marco Civil Law no. 12.965. (2014). *Chapter III, Section II. "Records, personal data and private communications protection"*. https://www.publicknowledge.org/assets/uploads/documents/APPROVED-MARCO-CIVIL-MAY-2014.pdf.

Ministério Público Institui Gabinete de Crise para a Copa do Mundo, Portal Brasil. (2014, 2 June), *Creative commons CC BY ND 3.0 Brasil*. http://www.brasil.gov.br/defesa-e-seguranca/2014/06/ministerio-publico-institui-gabinete-de-crise-para-a-copa-do-mundo.

Originally published 6 September 2013 by Agência Pública as part of its #SpyFiles3 special coverage. http://globalvoicesonline.org/2013/09/27/brazil-becomes-a-big-survelliance-market-player-for-the-world-cup/.

Östman, J. (2012). Information, expression, participation: How involvement in user- generated content relates to democratic engagement among young people. *New Media and Society, 14*(6), 1004–1021.

Pasolini, A. (2014). *Occupy: Brazil's lawmakers push anti-terrorism policy to stifle protest ahead of World Cup, Conectas direitos humanos, 6 March.* http://www.conectas.org/pt/acoes/midia/noticia/14984-occupy-brazil%E2%80%99s-lawmakers-push-anti-terrorism-policy-to-stifle-protest-ahead-of-world-cup.

Patry, M. (2014). *Brazil: A new global Internet referee?* Policy paper, Index on Censorship, p. 15. http://www.indexoncensorship.org/wp-content/uploads/2014/06/brazil-internet-freedom_web_en.pdf.

Police Raid Activists' Homes, Arrest 8 (anarchists, BB, antifa) On Eve Of FIFA's World Cup—Updates, Revolution News. (2014, 11 June). Courtesy Revolution News. http://revolution-news.com/breaking-police-raid-activists-homes-arrest-8-anarchists-bb-antifa-on-eve-of-fifas-world-cup-updates/.

Police violence during historic protest raises serious questions, Rio on Watch. (2013, 27 June). http://www.rioonwatch.org/?p=9797.

Reporters Without Borders, World Press Freedom Index, Reporters Sans Frontières. (2014). http://rsf.org/index2014/en-americas.php.

Rio de Janeiro inaugurates Command and Control Centre. (2013). Brazil Federal Government Portal da Copa, 4 June.

Schausteck de Almeida, B., Mezzadri, F., Marchi Jr, W. (2009). Condsideracoes Sociaisa e Simbolicas sobre sedes de Megaeventos Esportivos. MotrivivÍncia Ano XXI, no. 32–33 (June-December) pp. 178–192. https://periodicos.ufsc.br/index.php/motrivivencia/article/view/2175-8042.2009n32-33p178/14118.

Schmidt, B. (2014). Brazilian builder odebrecht emerges as World Cup Winner, Bloomberg Business News, 10 June 2014. http://www.bloomberg.com/news/articles/2014-06-10/brazilian-builder-odebrecht-emerges-as-world-cup-winner.

Teleco, Mobile Subscribers in Brazil (preliminary data report). (2014, 4 November). http://www.teleco.com.br/en/en_ncel.asp.

Waiselfisz, J. (2013). Mapa da Violência, Mortes matadas por armas d fogo, Centro Brasileiro de Estoudos Latino-Americanos/FLACSO, p. 11. http://mapadaviolencia.org.br/pdf2013/MapaViolencia2013_armas.pdf.

Wheeler, D., & Mintz, L. (2012). New media and political change: Lessons from internet users in Jordon, Egypt and Kuwait. In R. L. Fox & J. M. Ramos (Eds.), *IPolitics: Citizens, elections, and governing in the new media era* (pp. 259–288). New York: Cambridge University Press.

Willis, G., Muggah, R., Kosslyn J., Leusi F. (2013). Smarter policing: Tracking the influence of new information technology in Rio de Janeiro, Igarapé Institute report, November. http://igarape.org.br/wp-content/uploads/2013/10/Smarter_Policing_ing.pdf.

Universal Declaration of Human Rights, Article 19. http://www.un.org/en/documents/udhr/.

Ziccardi, G. (2013). *Resistance, liberation technology and human rights in the digital age* (pp. 17–18). Berlin: Springer.

From the Fringes: South Africa's Smart Township Citizens

Hylton Mitchell and Nancy Odendaal

Abstract This chapter explores the Information and Communication Technology (ICT) practices of the Social Justice Coalition (SJC), a community based organisation working with informal settlement communities in Khayelitsha, on the south-eastern fringes of Cape Town, South Africa. Many communities within Khayelitsha rely on communal toilets for sanitation, and the maintenance thereof has been highly uneven in quality. The chapter focuses on the use of ICTs in the SJC's advocacy work on the delivery of sanitation to informal settlement residents. Within the larger theme of citizen action, three main research areas were identified and explored: the appropriation of ICTs by the SJC's field staff, the use of their Web 2.0-based social networking sites (SNS) in their advocacy work and digital mapping of portable communal toilets. The research interrogates the use of ICT as a strategic tool for knowledge-based community empowerment, with the aim of understanding how these emerging uses of technology could assist developmental work within this sprawling area. The research emphasises the roles of various actors, the roles played by different technologies and the relations between technology and people. The emphasis on agency reveals that technology is not enough to augment empowerment processes. Technology–organisational relations are enmeshed within an institutional frame, where the meaningful use of technology requires a repositioning of network relations, in order for it to fulfil its potential as an empowerment tool.

H. Mitchell · N. Odendaal (✉)
University of Cape Town, Cape Town, South Africa
e-mail: nancy.odendaal@uct.ac.za

H. Mitchell
e-mail: hylton_mitchell@yahoo.com

M. Foth et al. (eds.), *Citizen's Right to the Digital City*,
DOI 10.1007/978-981-287-919-6_8

Keywords Information and communication technology (ICT) · South Africa · Townships · Empowerment · Khayelitsha · Cape Town · Social justice coalition · Communal toilets · Sanitation

1 Introduction

> Service delivery protests have developed some broad rules over the past three years, a formula to which the majority adheres. Regardless of how they start, they end in burnout. At some point, usually after three or so days, those who were manning the barricades, patrolling the streets and battling the police simply run out of steam. They never get what they want—because the houses or water infrastructure or other services they want could not be produced out of a hat, even with the best will in the world—and they rarely even get promised exactly what they want. They just stop.
>
> (De Wet, Mail and Guardian online, 23 March 2012)

Infrastructure in South Africa is inherently political. Under Apartheid, urban planners worked closely with engineers to create discreet townships serviced by the 'planned grids of urban life that were regular, clean, visible and open to surveillance…' (Minkley 1998: 210). Lack of basic services in black townships ignited social protest in the turbulent 1980s and to this day doggedly dominates the news, twenty years after the official end of racial segregation (since the first democratic elections in 1994). Infrastructure delivery remains core to the South African state's plans for economic growth and urban regeneration.

The chasm between grassroots aspirations and state intentions continues to be a stubborn feature of life in areas such as Khayelitsha, in Cape Town. While the country is considered a primary African economic hub, three South African cities top the list of the most unequal in the world (UN Habitat 2008). The institutional frame intended to enable delivery entails sophisticated bureaucratic machinery informed by a pro-rights constitution. Devolution of developmental autonomy to local government is intended to create the space for collaborative solutions under a fragile democracy. This chapter considers the extent to which digitally enabled citizen action can contribute to this goal.

This research considers the actions of a social movement in Cape Town, the Social Justice Coalition (SJC). The organisation uses mobile phones and social media to hold the City of Cape Town accountable to the residents of Khayelitsha with regard to the delivery of basic sanitation. Many residents rely on communal toilets for everyday use and the maintenance thereof has been a contentious issue. This piece explores the extent to which Information and communication technology (ICT) enables community members to exercise their constitutional right of access to basic services. There are three distinctive contextual elements worth highlighting here: a rights-based Constitution that advocates access to information and freedom of expression; decentralized local government with a firm developmental mandate; and finally, a messy and unresolved telecommunications policy environment that has had profound impacts on digital access. This chapter reflects on how this particular institutional context, together with a progressive constitutional environment coupled with socio-economic inequality, frames the environment for digitally enabled activism.

The continued evolution of ICTs merging with the urban context has gone from having a marginal influence, to becoming a central medium in the last decade. More literature tends to focus on the transformative potential of technology. What is more important, however, for the purposes of this research, is how the relationship between space, society and technology has impacted on urban governance processes. Media platforms, which O'Reilly dubbed Web 2.0 (2005: 6), are the products of innovation networks that emphasise user participation (Cha et al. 2009: 310). Digital content creators are now able to develop visionary and practical technology applications, and create urban services where communities can come to grips with the potential of a networked urban infrastructure, and make inputs in that regard. Web 2.0 influences how people communicate. As a result, information exchange becomes more fluid. In activism circles, political activists are able to now use these applications to create websites and social networking sites (SNS), where people are able to actively participate by gaining access to policies or legislation for example (Isaías et al. 2009: 118). Additionally, these platforms could be used in educating, mobilising and sharing information between participants as the interaction forms.

This chapter explores the extent to which this is achieved in Cape Town, by first, providing an overview of the evolution of social movements in the information age in South Africa as well as the limitations in terms of technology access. Three dimensions of how the organisation uses technology for its activist programme are explored. The final section analyses the SJC's activities in relation to some of the elements of a digital democracy: information flow, the roles of technologies in enabling information flow and distribution, the development of new networks and the development of local capacity in making inputs into local governance processes.

2 Social Movements in the Information Age

The theme of this research is to explore the use of ICTs by civil society and digital activist communities as inputs for creating public accountability and a participatory form of government. It is generally accepted that civil society constitutes a political power block and could contribute to a platform for real public engagement, where non-state actors can shape debate in the public sphere, influence opinions, and foster social change. Subsequently, social media technologies have allowed for participation in a new type of public sphere that acts as an alternative to mainstream media. The result is that digital technologies can now direct communication between citizens and those in political power (Van Dijk 2000).

In the early days of the network society debates, Castells maintained that Internet networks provide an independent space for movements to form and overcome geographical and social boundaries (2004: 183). Although the precise role of the digital networks in the Arab Spring, for example, is still being debated in various academic circles, Castells goes on to explain that the Internet and mobile phone networks contribute to political autonomy (2012: 11). Howard finds that ICTs strengthens democracy and increases civic involvement and autonomy of civil society and in the context of the Arab Spring, challenges dictatorships (2011: 108). The influence of ICT has also been shaped by the fusion of globalisation, increased online connectivity and knowledge networks. Howard and Hussain found that the extensive use of digital networks by the youth predominately had a significant impact on the intensity and power of demonstrations during the Arab Spring (2012: 49). They go on to point out that Facebook and Twitter did not cause the revolutions, but that the careful and strategic use of digital media to network regional publics and international support had empowered activists in new ways that go beyond online activism.

South African social movements continue to use the new ICT revolution to enhance participation, create access to information, deliver alternative content stream and exchange information in digital forms, substantially narrowing the digital gap common in developing countries. However, other factors to consider in this context include the sluggish transformation of the ICT policy landscape in South Africa and the African continent as a whole. These issues impact ICT's effectiveness as an advocacy tool for social movements to disadvantaged segments of society.

In South Africa, municipal service–delivery protests are often organised responses to local government delivery failures that manifest in depressed urban communities. But, these manifestations can also be seen as the result of a failure to integrate digital networks on the side of marginalized citizens to voice their discontent with the powers that be, and to mobilise around protests by inducing 'flash mobilisation' (Rheingold 2002: 48). There has been a tendency to bracket urban movements with local, municipally orientated protests and community activism, which are closer to the character of organised interests than the broad, value-led, socially transformative collective movements more typically associated with 'new social movements' (Hamel et al. 2003: 167). Collective political action in cities takes a variety of forms, and it is organisationally structured according to the nature of the goals and objectives that groups set themselves, framed by the opportunities and constraints provided by state actors and the resources and obstacles within civil society, and in particular by the nature of the political field in which parties, interest groups and movements have to operate within the urban/regional context (Parker and Song 2006). Whatever the form that political action assumes in cities, communities remain key interlocutors and agents of urban change. Therefore, embracing and integrating ICT as a tool presents enormous challenges in the process of transforming the South African NGO sector as a whole. The challenges present themselves at two levels: first, at the level of raising awareness, and secondly, at the level of implementation (Rahman 2005: 17). Seemingly mundane issues that relate to access and infrastructural efficacy continue to inform the context for digital activism in South Africa.

The South African government's approach to ICT roll-out mimics international trends in that it aims to reduce the access gap of poor citizens and promote technology as a stimulant for economic growth. Reducing income disparities, high unemployment rates, inequality and poverty have been some of the greatest challenges of a post-apartheid South Africa. While significant advances have been made when it comes to access to ICT technologies, with South Africa's telecommunication system described as one of the best on the African continent, Internet services reached less than 3 million users in urban centres in the early 2000s (Maldonado et al. 2006). Last-mile programmes that target digital access in community facilities (such as the Smart Cape programme in Cape Town; see City of Cape Town 2015) and the broadened availability of smart phones have contributed to some growth in Internet access. World Wide Worx reports that the South African Internet user base had grown by 25 % from 2010 to the end of 2011 (2012). Overall Internet access remains low, however. In Cape Town, for example, the telecommunications profile of the city shows high access to mobile telephony (93 % of the population has access to a cell phone), but very low computer access (37.9 %) and even a lower rate of access to telephone landlines (34 %). Just over 50 % of the city's population has no access to the Internet (StatsSA 2011).

Moodley (2006) argues that differential access to ICTs is linked to other social divides such as income, education, age, gender and ethnicity. Infrastructure roll-out is not enough. Limited access to the Internet has a racial configuration that classifies the digital divide even further, as the domain of a 'privileged minority' to engage in political debates and discussions. However, government policy acknowledges the existence of historically disadvantaged communities and encourages the adoption of telecommunication services by these groups (Maldonado et al. 2006: 143). Latchem and Walker (2001) argue that telecentres have proven to have the greatest potential to narrow the 'digital divide' in remote, rural and other disadvantaged areas. Singh (2010) critiques this model arguing that the political economy of ICTs in South Africa is influenced by bureaucratic incompetence, contradictions and incoherence in government policy and government unable to regulate pricing. In addition, other more pressing government priorities such as healthcare, education, rural and economic development and basic service delivery remain more challenging priorities in the South African sociopolitical context.

Literacy levels and the dominance of English are other contributing factors in preserving the digital divide domain in the hands of 'the haves' which makes the democratic nature of the Internet and social media still an unfolding reality in South African and the continent. Without structural government and private-sector support, the 'digital divide' will continue to manifest itself along socio-economic dividing lines, reinforcing the already existing social inequalities. However, government policy and mobile technology changes plus infrastructure development have proven to be the method with the greatest potential to narrow the 'digital divide' in Africa. The question remains how these various dimensions of the digital divide influence the actions and potential of social movements in South Africa. The following case provides an unraveling of these issues.

3 The Scope of Social Justice Coalition's 'Slow (Digital) Activism'

> It is time that the City of Cape Town and local authorities throughout the country get used to the idea that the urban poor do not need to wait for experts to provide data on conditions of life in their communities.
>
> (Robins, in The Cape Times, December 15 2014)

Cape Town is the second largest city in South Africa (after Johannesburg), with a population of just over 3.7 million. Its regional GDP per capita compares to that of Mexico City and Naples, with a value of US$15.250 per capita, 40 % more than the national average (OECD 2008). Outside the Gauteng City Region (that includes Johannesburg and Pretoria), it is the only major city region that has increased its share of national output since the demise of apartheid. Its economic base has shifted to that of a service-based economy, representing 69 % of total regional GDP and employment. The tourism and hospitality industry is a big part of this growth. The creative and knowledge intensive industry has contributed more substantially recently (OECD 2008). Despite these positive macro-indicators, the city suffers from serious inequalities. Only 76 % of the labour force is employed, with just over a third (35.7 %) of households earning an income of less than US$300 per month (City of Cape Town SDI and GIS 2013). The growth in the service sector does not necessarily benefit the low- and medium-skilled population living on the outskirts of the city.

The city's population had increased by almost 30 % since 2001, with a large number of migrants from other parts of the country living in informal settlements on its south-eastern fringes. Fourteen percent of households live in informal dwellings in informal settlements; there has been a large growth in the number of households living in informal dwellings in the backyards of formal homes also. Seventy-eight of the city's population is formally housed. With informal living conditions come disparities in service infrastructure access: 3.7 % of households do not have access to electricity for lighting; 8.8 % have no access to sanitation onsite, and there are 232,027 households registered as indigent (City of Cape Town 2015). Service access is highly differentiated spatially. In Khayelitsha, home to 391,749 people in the south-east of the city, only 45 % of households live in formal dwellings, 62 % of households have access to piped water in their dwelling or inside their yard, and 72 % of households have access to a flush toilet connected to the public sewer system (City of Cape Town SDI and GIS 2013).

The sanitation situation has been one of the most controversial service delivery issues in recent years. While the number of households with access to a flush toilet connected to a sewerage system increased from 85 % in 2001 to 88 % in 2011, the number of households using bucket toilets has increased from 34,200 in 2001 to 48,500 in 2011(City of Cape Town SDI and GIS 2013). Inadequate maintenance of communal toilets, the local government's supply of temporary rather than permanent facilities and continued implementation of the bucket system have sparked

Fig. 1 Flush toilets—RR Section. *Source* Groundup Media

what is now termed 'poo protests' in the city (Cape Times 2014) with individuals flinging human faeces collected in bucket toilets on one of the national freeways into the city (Cape Times 2013b), and depositing human waste in the foyer of its international airport (Cape Times 2013a). Other efforts at activism reflect the more systematic approach of the SJC (Fig. 1).

The SJC Cape Town is a civil society organisation based in Khayelitsha, with 11 active branches and 40 partner organisations. The SJC with its branches coordinates two primary campaigns, namely 'The Clean and Safe Sanitation' campaign and 'The Justice and Safety for All' campaign, through education, policy engagement and community mobilisation. It does this through its onsite presence in Khayelitsha and other locations, but also through social media. The organisation's website (www.sjc.org.za) includes online petitions, responses to public press releases and links to news articles. YouTube is used to augment campaigns. Twitter and Facebook are used to stay with the facts in real time as on the ground information is shared in among press monitoring and meeting announcements. The website includes donation information in South Africa and the USA. Collaborative campaigns can be accessed via the site such as the 'Campaign for Safe Communities' that includes an online petition and multimedia evidence of issues campaigned against.

In a recent op-ed piece, Robins (2014) provides a nuanced uncovering of the organisation's approach: data-driven mobilisation where information is used to empower locals in order to confront city government on service delivery. This research is a systematic examination of the mechanics and elements that enable SJC's 'quiet activism'.

Fig. 2 Scope of fieldwork and analytical frame

The fieldwork focused on three dimensions of the SJC's work. The aim was to interrogate how each of these interrelates with stakeholders and evaluate whether ICTs can assist in knowledge-based community empowerment (see Fig. 2). The first dimension refers to how the organisation applies knowledge-based systems in managing its relations with the City of Cape Town and beyond. It interrogates the organisation's online social networking indicators and ICT platforms. The second dimension relates to how the organisation functions within community networks, and how SJC activists function in the field and in their office. Finally, the third focuses on the technological features of a digital mapping system of communal toilets in an area called RR Section in Khayelitsha are examined as a microcosm of how information is retrieved and relayed.

In the analysis section that follows the presentation of the research findings below, questions are asked of the direction of information flow, the extent to which new networks are enabled through technology-augmented activism and the uptake and influence of different media on the organisation's activities (Fig. 2).

Data were collected over two months in 2014. The data collection techniques included questionnaires, in-depth interviews, focus groups and direct observations within the organisation and monitoring their ICT activities *in situ*. Each of the three dimensions of the SJC's activities is reported on below, starting with its community networking, followed by its online activism and finally, its nascent digital mapping programme.

3.1 SJC Fieldworkers and Community Networking

Through a series of semi-structured interviews, fieldworkers were probed on their use of digital media and/or mobile phone platforms in their fieldwork. Here, the aim was to develop insights into their ICT practices and how ICTs played any part in their fieldwork around the theme of 'decent sanitation' provision and grassroot mobilisation. Seven SJC fieldworkers were interviewed; participants were all black South Africans, two were female and five male, averaging between 21 and 25 years of age.

The primary role of the SJC fieldworker was to service community residents with training workshops, capacity building, project management and raising awareness on sanitation campaigns across the informal settlement community. Fieldworkers were mandated to investigate, support and advocate on behalf of informal settlement residents around 'decent sanitation' delivery in Khayelitsha. This they did by improving the flow of strategic information and communication of participants in informal knowledge networks.

Only 4 out of 7 SJC fieldworkers had the type of smart mobile phones that could access the Internet, social media and their personal email accounts. Frequent use of their mobile phones was limited due to high Internet data costs, theft and other safety concerns that existed in the township. In addition, any digital recordings or videos documenting sanitation delivery failures were subject to SJC office staff scrutiny and verification before it could be uploaded to any of the SJC social media networks such as Facebook (facebook.com/pages/Social-Justice-Coalition) and Twitter (@sjcoalition). Fieldworkers as a result had no direct access to the SJC online presence as SJC office staff confidence were low in their ability to curate uploaded field content due to fieldworker's lack of proper media training and skills needed to execute these tasks. This resulted in only one or two SJC office staff having official access to the SJC SNS, which resulted in an ad hoc displacement of field information online.

Additionally, respondents explained that few informal settlement residents used the Internet as a rights-based activism tool but rather used it primarily for entertainment, connecting with friends and socialising. Based on fieldworker interactions with residents, Internet penetration was very low as older residents were sceptical of digital recordings or placing responses on video record. This further hampered an integrated acceptance model for digital community knowledge building as residents preferred traditional media such as radio or the printed material. The use of new media was relatively an unfamiliar practice due to a lack of digital skills, appropriate handset devices and general availability of digital infrastructure and Internet access in deprived communities. However, respondents maintained that this aversion to digital technology was due to the generational gap among older residents, as young people were familiar with the Internet and were frequent users of the few local Internet cafés in the area.

Results from the respondents' online digital access portals were illustrated to explain their average weekly online activities and online activity networks, which would increase the understanding and provide an overview of SJC community fieldworker constraints (Table 1).

Table 1 Online access per week

SJC community advocates	Category	Facebook access per week (h)	Twitter access per week (h)	Internet access per week (h)
Desktop Internet access	5 Males	1	0	1
	2 Females	2	0	1
Mobile Internet access	5 Males	3	2	1
	2 Females	4	2	1

3.2 SJC Office Knowledge and Social Networks

The semi-structured interviews were supplemented with in-depth interviews with the 5 (4 males, 1 female) SJC office staff. The aim was to highlight key features in their approaches to extending ICT-based community social networks and knowledge sharing processes by monitoring content flows on the SJC's online campaigns. The in-depth interview method allowed the research to gain the most tangible insight into the SJCs ICT practices to ascertain how the online social network architecture developed into the temporal spark—to understand how agency was formed. Wellman (1996) maintains that ICT-based community networks can lead to the development and maintenance of strong, intermediate and weak ties that are able to provide support and information across both broad and specific relationship within a community.

Studies in the community informatics (CI) field have delivered results in how ICT-based participative practices can assist a community's socio-economic and cultural goals. The aim here was to identify within the SJC's online content flows, opportunities to effectively integrate organic network of knowledge sharing between SJC stakeholders online. The research therefore focused on the information flows underlying such campaigns, examining the total number of messages accessible on the social media platforms, average message content and the frequency plus the reach of SJC content on these networks (see Table 2). The social media platforms used by SJC were Facebook, Twitter, and WhatsApp. The SJC website was the primary source of information (with associated links to YouTube and the aforementioned social media). In addition, the research developed data sets in which the SJC participants chose to constitute an online community through language, content categories and external community feedback.

Observation of online activity enabled an understanding of how information and knowledge were disseminated to a broader audience and references to the City of Cape Town. This included the perusal of the 38 SJC Facebook messages posted on their digital media platforms between May and June 2014, as well as monitoring the types of connections generated over the two-month period. Respondents explained that SJC Facebook knowledge distribution was not considered a priority, based on limited media training and lack of dedicated personnel in monitoring

Table 2 SJC Facebook content and interactions averages per week

Dimension	Average	Descriptive	Informative	Interactive	Civic message
Story output—posts per week	3–4 posts per week	Policing, sanitation and safety	Access to SJC local info and city information	No calls to action is related in the posts	Viewers gained a better understanding of the issues
Story content 'Like' per posts	0–3 'Likes' per week	Lack of user engagement	Little or no feedback user community	No calls to action is related in the posts	Viewers gained a better understanding of the issues
Participants replies per story	0–2 users reply to stories	Lack of user engagement	Little or no feedback user community	No calls to action is related in the posts	Viewers gained a better understanding of the issues
Participants comments per story	0–2 users reply to stories	Lack of user engagement	Comments in support of SJC actions	No calls to action is related in the posts	Viewers gained a better understanding of the issues
Other online content sources	4 posts from other content streams	News, media articles and other NGO support	Policing, sanitation and safety messages	No calls to action is related in the posts	Viewers gained a better understanding of the issues

and managing online engagement levels. As a result *'digital activism'* was treated as an 'after-thought' in its stakeholder mobilisation and network-building efforts. According to SJC office activists, they had developed an over-reliance on face-to-face and traditional 'grass-roots' approaches to implementing SJC operational functions, mainly because of the contextual environment they were operating in. Respondents, however, understood the potential value of effective online use of ICT among community networks, but this was impeded by operational and the lack of internal media skills necessary to optimise the inclusive and accessible nature of these ICT networks.

At the time of the research, the SJC Facebook page 'likes' stood at 2163. From detached observation of the SJC's online campaign messaging on Facebook, content was sent out in an ad hoc manner and contained an average of 3/4 posts per week during the time of the research. From the two-month observation period, 90 % of all sanitation-related information, stories and content on Facebook were SJC sanitation campaign articles or activism stories that were exported or lifted from other news sources such as the *Mail and Guardian* newspaper website; another was an online publication called *Groundup News*, sister organisations such as the *Equal Education* and the SJC official website. The Facebook posts were all in English.

3.3 Mapping in RR Section

During the research period, the SJC was in the process of rolling out a digital location-based mapping system that would map public portable communal toilet locations in RR Section, a largely informal area consisting of about 300 (1200 people) households in Khayelitsha (www.sjc.org). The intention was to create an interactive website and phone-based reporting system that could improve the efficiency in toilet maintenance in this area. The various methods used in mapping portable communal toilet conditions in RR Section of Khayelitsha provided the most local scale interrogation of the SJC's activities (Fig. 3).

Previous SJC efforts in assessing this were limited to sampling from a specific number of community residents in order to investigate system failures in particular. This mapping system allowed staff members who previously struggled with the accuracy of the data, to plot comparative reports in real time using mobile phones to monitor the maintenance of toilets or take photographs. Data collection methods employed were observations, semi-structured interviews and tracking the distribution of collected information online. Uses of data mapping of the sanitations services, portable toilets and the use of locations based software were recorded.

The research revealed important spatial qualities; most of these portable toilets were located on the edge of the RR Section informal settlement as inaccessible roads and high-density shack dwellings made it difficult for sanitation service providers to deliver portable toilets close to some residential structures. In future, the mapping system would make the need for SJC field/community advocates to conduct monthly audits obsolete, as resident feedback and input would be in real time. The maintenance of service delivery would now rest with local residents as

Fig. 3 Aerial map of RR Section. *Source* Groundup Media

they could report problems with portable toilets in real time, and input complaints directly into the digital system. At the time of research, however, the data collection process was still being done manually as the system was not fully operational. The monitoring and data collection process was still in an analogue stage as face-to-face interactions with residents on the ground to record digital numbers and profiling complainants were still being done with SJC fieldwork staff. The SJC has envisaged this portable toilet monitoring system becoming fully digitally integrated but the technology platforms and an organisational partnership plus funding needed to operationalize this system still needs more attention. It is envisaged that these location-based monitoring systems increase a bottom-up approach to community participation central to SJC engagement objectives.

4 Understanding the Limitations and Potentials of Digital Activism

The research findings, discussed above, indicate that the SJC's ICT-based community empowerment initiatives evolved in a particular economic, social and political setting. The SJC has made a concerted effort to remove internal barriers to increase community involvement and to strengthen democratic participation. However, it resulted in a limited evolution of an ICT-based community network. In order to explore this more thoroughly, this section analyses the extent to which the three dimensions of the SJC's work enable distributed information flow, what the roles of various actors and technologies are and how those articulate with the intentions of mobilisation.

4.1 Information Flow and Distribution

The SJC's online activity fits three categories:

- Informative Messaging—Content that conveyed digital material that could stand on its own.
- Interactive Messaging—Content that required a response from other digital participants.
- Civic Messaging—Content that came from online community participants.

A recent skirmish with the City of Cape Town provides a snapshot of the ways through which SJC employs social media to represent and disseminate information. The organisation's social audit of four informal settlements revealed that the city administration's janitorial system was not working (SJC 2014). The report is available online, with an associated YouTube version, with all associated maps and spreadsheets also downloadable. The City of Cape Town's response to the report was defensive and included a claim that SJC was using social media platforms to embarrass the city administration (Robins 2014). The SJC responded accordingly, online, supported by the data necessary to augment its position. The transparency of such exchanges, and the painstaking attention to operational detail and meticulous description of data collection practices speak to community insight and data literacy, a form or empowerment that reveals what Robins calls 'governance from below'.

The online dissemination of surveys and reports, as well as links to media from activist organisations, as a counter to municipal evaluations has proven to be one of the central tasks of the organisation. Using this information to motivate for more rigorous upkeep of communal toilets has resulted in revising service level agreements between the City of Cape Town and contractors. The fact that these audits on sanitation services are updated monthly is central to ongoing operations.

Social media are used in day-to-day monitoring. One of the drop-down menus on the website is entitled 'Imali Yethu' a Xhosa phrase for 'our money'. The menu contains pages with places where service delivery contracts are in place for sanitation; refuse removal and policing in Khayelitsha. The detail of each contract is displayed, with the expected frequency and scope of maintenance tasks, the contact person and detail of the company contracted and time frames. The aim was to empower communities with the necessary facts to monitor and share information on service delivery. Thus, information is captured offline, through use of cell phone cameras and/or note taking, and then uploaded at the SJC office. This does lead to a centralisation of functions, given that a small number of SJC staff has access to online media, leading to some operational difficulties and time lapses.

Online content tended to be one-directional, functioning as an information portal and also serving as an information resource in placing pressure on the municipality to maintain toilet services. The overall aim was to increase monitoring of public spending and accountability. Each page contains the usual Twitter, Facebook and other social media sharing facilities. As a mobilisation strategy, more broadly, SJC's online activity is constrained by limited Internet penetration in Khayelitsha, the relatively high cost of smart phones and entrenched preferences for face-to-face contact and simply talking on the phone. Using radio is preferable to the Internet, which in many ways is seen as unfamiliar, although the local application 'MixIt' (a well established South African forerunner to 'WhatsApp') is used for two-way communication between SJC staff and core community activists.

The efficacy of the SJC online strategies appears to be vested in its ability to respond timeously to issues using social media and its physical proximity to communities in need. Its mobilisation function is mainly through partnerships with other organisations such as alternative news source 'GroundUp Media'. Another function is engagement with citywide issues in a considered and informed way. This can be illustrated in a forum (organised together with the University of Cape Town- based African Centre for Cities) interrogating the potential for design to become an instrument of social justice and urban inclusion in anticipation of the designation of the city as World Design Capital in 2014. The event included inputs from academics, designers and activists at various sites around the city. Its symbolic value as an online and onsite space to interrogate the creative city discourse provided a meaningful counter to the usual notions of innovation.

Perhaps the most profound innovation to support bottom-up action is the virtual mapping system, designed to provide a historical data map and by making localised and therefore critical contribution to community-generated information.

4.2 New and Existing Social Networks

The fieldworkers were operating in a challenging environment with contextual constraints hampering the opportunity for ICT-based network building. The absence of network alignment was due to informal settlement residents having low Internet access, lower literacy rates, poor infrastructure, high poverty levels and technology phobias. The fieldworkers as the macro-actor in this instance wanted to employ mobile phone digital application, digital networks and SNSs as viable avenues to collect, store and disseminate informal settlement sanitation information but socio-economic conditions ultimately altered how they used the technology. Aspirations towards increased awareness, digital documentary proof and build stronger participative nodes within the community were voiced. They additionally believed the mediated digital platforms could reduce the information gap and in turn speed up the process in holding the municipality accountable for sanitation failures.

Social media and mobile technologies have become integral parts of the SJC's organising and work practices. In addition to influencing and increasing the way they engaged with citizens, outside stakeholders and communities, the SJC understood the medium as a valuable participatory tool. A large number of theorists have linked the development of large-scale organisations and other social structures to developments in information technology (Bell 1974; Beniger 1990; Castells 2004). According to Weare (2002), the broadcast capability of the Internet may allow information to be shared throughout a group efficiently, and thus reassure members that they are on an equal footing where information access is concerned.

But online participation was limited for informal settlement residents and subsequently, the SJC office activists through their online engagement identified outside stakeholders, such as the City of Cape Town, sister NGO organisation, such as *Equal Education* and *Ground-Up* news, and news sources such as the *Mail and Guardian* plus SNS users on SJC digital platforms as relevant actors to foster their urban agendas. At a rudimentary level, the SJC sent out longer messaging overall on sanitation campaigning, which was accompanied with pictures on sanitation, related issues concerning the Khayelitsha community. This digital communication channel allowed the SJC to quickly share content that could be made available to all users immediately. Facebook allows for immediate feedback by being able to 'reply' to content posts or users are able to posts content on the SJC Facebook 'wall'. According to Ferebee and Davis, what makes Facebook a central communication medium is its open-source nature that allowed rapid communication, rapid change, and rapid development (2009). Users who formed part of the SJC Facebook community could 'share' and 'like' the message feeds. Ferebee and Davis call these 'interaction points' on Facebook (2009). This allows the whole online community to participate, as message posts and news feeds would automatically reach all connected users from the Facebook group. Posts 'likes' during the research period were on average 1–3 'likes' per posts/content message, out of a SJC Facebook user population of 2163 connected community users. Facebook is unique as a platform as content producers have access to the online response data generated on story posts and the amount of users who have interacted on the page. The user community experience is enhanced based on the automation and architecture of the digital communication platform. The observation showed that the message posts 'likes' were extremely low at 1 and 3 per story posts. The response rate might be low but these Web 2.0 platforms facilitate people in collaborating and sharing ideas and content.

As the main actor in the network, the SJC was aligning its content with outside stakeholders' needs with the intention of addressing its informal settlement agenda. Computer-mediated communication facilitated the proliferation of interest group activities. SNS were moderately successful in mobilising parties based on similar interest and their ability to influence public opinion by sharing public accounts of communities in sanitation challenges based on articles and covered incidents. Interest groups formed part of the actor-network. SJC online sanitation campaigns did influence opportunities for diverse community interest groups outside Khayelitsha to participate on a range of SJC campaign-related activities, events and notices. In the case of both the SJC and other interest groups, the research revealed that new technologies enhanced network building, suggesting that it could aid community development.

4.3 Broader City Governance Impacts

During the research period, there were 112 Msengu toilets (portable toilets) identified and located as problematic in RR Section, according to SJC fieldworkers, this translates into one toilet for every 3 families in this location. This information was duly publicised on the website and used, together with other audits to confront the City of Cape Town. The feedback received from informal settlement residents by SJC community advocates was that such processes had started to influence how service level agreements were being maintained between contractors and the City of Cape Town.

Some of the issues are specific to subcontractors, and duly brought to the attention of the municipality. For example, the manager of one of the companies which supervises the janitorial cleaning had not been paid in some instances—as based on user feedback and monthly audits. The SJC used these monthly audit reports on sanitation services as information to inform the City of Cape Town.

The organisation also claims to have assisted the City of Cape Town in adapting planning and roll-out strategies more effectively for sanitation provision in informal settlement areas in Khayelitsha. The intention of the collaborative mapping initiative would allow SJC monitors to identify patterns of where toilets are not being properly maintained and help them recognise where issues were more prevalent. Such spatial data-driven activism is intended to assist the City with maintenance information while also enable monitoring by community members.

4.4 Roles of Tools and Technologies in Networking

The SJC's computer-mediated communication enabled interested parties to communicate on the issues raised in various campaigns and events taking place in the informal settlements. This was also with face-to-face contact in campaign activities. It does appear from SNS 'likes' and followers that the SJC had managed to attract mediated support on its platforms. By using the technology to facilitate dispersion, the SJC was able to reach a broader community, extending its relevance in time and space. SJC staff preferred using their personal Twitter accounts to engage in dialogue with certain political figures. However, the SNS platforms did not necessarily lead to more engagement with informal settlement residents.

SNS networks transcend the spatial community, thus, they link people with similar interests, by forging links among people sharply different from one another. Twitter and Facebook as social media platforms can be described as computer-mediated communication among large groups of individuals. These groups of people might not meet face to face to exchange content, ideas and experiences. These digital interactions in which people do not have to know each other are making new kinds of communities possible.

4.5 Development of Local Capacity

There are three major groups of actors involved in the SJC online Facebook campaign actions; the SJC office staff, intermediary sources supplying online content on sanitation issues and the SJC online users who took part in online activism campaigns. Each performs a distinct action online in activating campaign messaging. Although their actions are coordinated to serve the SJC's online activism agenda, the differences between action and responses had implications in measuring its effectiveness. Online content were one-directional in nature and seldom had a call to action response leading to low or no direct involvement from user responses in SJC online campaign actions. Initial staff responses were that SJC did not have a particular target audience in mind as part of its online distribution model.

The SJC operates in an environment in which digital communication is not strongly promoted within the organisation's daily activities. This led to greater centralisation of communication within the organisation, as only two senior office members became the conduits through which messages would pass. This can be attributed to a number of factors such as insufficient media skills of general office personnel, the centrality of communication infrastructure in the organisation and the external environment in which the SJC operates which has low Internet penetration.

According to SJC office staff, because of low Internet penetration in informal settlements, it is more advisable to conduct SJC campaign issues on the phone or talk face to face with people. Community residents are more amenable when you share information face to face and use radio slots as a means of communication as the Internet still remains an unfamiliar and uncertain communication terrain.

The analysis of the relations between users could indicate the degree of 'transitivity', which could be interpreted, as indicators of stability. Further, ICT characteristics such as connectivity and its one to many configurations could potentially have positive effects on increased community participation. Because when we consider the community as a node within the communication network, their roles as nodes stem only from their ability to contribute to the 'network goals' (Castells 2004). As Monge and Contractor explain, 'communication networks are the patterns of contact that are created by flows of messages among communicators through time and space', (2003: 39). Therefore, the high reliance on face-to-face interaction in completing the sanitation audits by SJC community advocates disrupts the network process flows. This can also be attributed to informal settlement community's general distrust of digital-mediated media, lack of regular Internet access, affordability of Web 2.0 enabled smart phones, low technical skills levels and other socio-economic barriers.

Strengthening SJC's information/digital infrastructure and developing basic digital communication channels internally would alleviate the difficulty in building a case or taking on the ground activations in the digital sphere. The main goal would be to strengthen the SJC's internal digital communication infrastructure. It should include a campaign analysis of SJC on the ground activations and how to make them better known to local government decision-makers, local authority planners and other forum participants who are urban transformation experts. Here, the focus should be on developing digital communication channels that could create greater engagement and connectivity, creating an enabling environment to speed up service delivery bottlenecks and infrastructural development, in a real-time and integrated digital communication environment.

In addition, ICTs have a greater potential for improved engagement on sanitation challenges if it involves the township community on these digital platforms. SJC needs an open and accessible digital channel for township citizens to express their views. A strong emphasis should be on the appropriation by social practice, which seems to be based on SJC community advocate feedback heavily concentrated among teenagers. The SJC should consider new forms of community, bringing young people together online around shared values and interests, that can create ties of campaign support that already exist in on the ground service delivery protests and to extend this to online interactions.

The socio-economic status of citizens explains their level of online participation in terms of the social circumstances that shape their attitude to ICTs. In one of the most widely employed online participation theories, Syed and Ali (2010) argue for consideration of various resources, motivations and civic skills as antecedents of online participation. So, how can digitally mediated communication facilitate community empowerment? First, the maintenance and management of dispersed face-to-face networks can be dramatically increased. It can facilitate a larger assessment pool of community actors creating ever-growing networks of allied participants. Digital platforms such as SNS can transform displaced communities into digital networks, where notices can be placed, and allow community actors to stay in touch beyond face-to-face meetings and exchange text and data. Certain mobile phone digital applications will allow community advocates to form closed and direct virtual networks with their constituents to facilitate mediating service delivery failures in a more effective manner. The virtual community could also become an authentic community experience for poor informal settlement residents if it was based on the everyday physical world experience of community actors. Community activists associations that seek to strengthen their communities require cohesion among their constituents or members and infrastructure that encourages participation.

5 Conclusion

The research explored the extent to which ICT enables community members to exercise their constitutional right of access to basic services in Cape Town. The research probed the extent to which digital linkages impact on the relationships between the communities and the state agencies that are accountable for service delivery. What emerges is that digital-enabled activism, in this case, has not yet fulfilled its potential of a two-way collaborative form of democracy. Digital tools have been effective here in monitoring a particular form of service delivery and holding local government to account. This is an important part of local governance, and indications are that this 'quiet activism' could become more entrenched as local capacities develop.

Community activism can never just be digital; technology coexists with other forms of communication. The networking functions of the SJC are important inputs into a broader exchange around inclusive cities. Use of technology does not automatically translate itself into a radical online participation tool for continuous citizen engagement, but part of a suite of tools. These actions are not enough poor service delivery per se; or to radically empower the poor, the digitally disenfranchised, and dispersed organisations against the powerful. Structural and institutional issues are ongoing considerations while computer-mediated communication could potentially reinforce the existing power structures within society.

These results serve as a novel framework, in whether the experience of computer-mediated communication encourages a detailed understanding of the structures of power that lie behind machines, the Web and of disadvantaged communities. Proponents of the Internet have long advocated a kind of cyber-democracy and the civic-led participation that would drive development and community empowerment. ICTs have the potential to become important catalysts for interaction and creating community participation within close-knit informal settlement communities. But, clearly virtual communities as set-up by organisations such as the SJC with low technical skills levels, do not flourish on close inspection but only promote a centralised experience, as community participants are largely marginalised in this gravitation towards online activism.

Certainly, this ICT platform could engender profound changes to communities. Further research is needed to examine the degree to which digital systems can enable two-way flows in information production and delivery at varying scales. The potential for mapping systems in making cities more accessible and livable for marginalised communities is immense. In addition, the causal relationship, between ICT and how it influences what a community knows and understands about their developmental needs, needs further interrogation. ICTs change working methods and the ways in which people communicate. Similarly, it affects how the community accesses and shares it with others and establishes information as source of power (Heeks 2001). ICTs can potentially create an ecosystem through which poor communities can directly help improve the welfare of their urban context. The concern is how modern ICTs can be utilised to strengthen and develop the information systems for marginal communities. However, recent trends of information and communication technologies are rapidly consolidating global communication networks with implications for communities in developing countries. The implications and constraints of existing information systems on poor marginal communities and their intersection with ICTs are also little understood in relation to urban transformation. The SJC case shows us that the potential for a more finely grained urbanity, connected into broader networks, exists.

In the end, the research had delivered some practical insights and implications, which showed that ICT actors could play a major role in the formation and extension of community empowerment and foster participation by enrolling various online campaigns. For example, the Internet influenced communication, information, collaboration and knowledge management in an urban environment not traditionally associated with technological innovation. Some online users had sufficient incentive to participate and some users distributed and shared voluntarily the SJC content of their campaigns online. As the opportunity costs associated with mobile Internet access decrease, the promising signs are that this will empower marginalised groups.

The use of embodied infrastructure, such as cell phones and associated applications, could bring poor urban communities into the information age, with the possibility of generating bottom-up knowledge and making inputs into governance processes. Local government stakeholders, urban planners and policy makers need to acknowledge information poverty as a form of poverty. ICT infrastructure provision should form part of the essential service delivered to communities is the goal of empowerment is to be realised. Only then will cities such as Cape Town be able to fully develop into digital cities. This case shows that the potential exists for collaborative, media-enabled governance to be considered a part of contemporary urbanity in South Africa and similar cities elsewhere.

References

Bansal, N., & Pang, A. (2010). New media: a new medium in escalating crises? *Corporate Communications: An International Journal*, 15(2).

Bell, D. (1974). *The coming of post-industrial society: A venture in social forecasting*. London: Heinemann.

Benkler, Y. (2006). *The wealth of networks: How social production transforms markets and freedom*. New Haven: Yale University Press.

Beniger, J. R. (1990). Conceptualizing information technology as organization and organizations and communication technology (p. 29).

Berawi, M. A. (2004). Quality revolution: Leading the innovation and competitive advantages. *International Journal of Quality and Reliability Management, 21*(4), 425–438.

Cape Times. (2013a). Poo protests: 'It can't carry on' (online). Nontando Mposo. Cape Town, August 8. Available from http://www.iol.co.za/news/politics/poo-protests-it-can-t-carry-on-1.1559618#.VWSGD1lViko. Accessed 24 January 2015.

Cape Times. (2013b). Faeces fly at Cape Town Airport (online). Zodidi Dano and Clayton Barnes. Cape Town, June 26. Available from http://www.iol.co.za/news/crime-courts/faeces-fly-at-cape-town-airport-1.1537561#.VWSGxVlViko. Accessed 24 January 2015.

Cape Times. (2014). Data-driven activism empowering (online). Steve Robins. Cape Town, December 15. Available from http://www.iol.co.za/capetimes/data-driven-activism-empowering-1.1795657#.VWSB0FlViko. Accessed 24 January 2015.

Castells, M. (2000). The rise of the network society: The information age: Economy, society and culture (Vol. 1). Chichester: Blackwell.

Castells, M. (2004). *The network society, a cross-cultural perspective*. Massachusetts: Edward Elgar.

Castells, M. (2012). *Networks of outage and hope, social movements in the internet age*. Cambridge: Polity Press.

Cha, J.-W., Lee, H., Han Y.-S., & Kim L. (2009). Changwon National University, Changwon, Republic of Korea, Department of Computer Science, Yonsei University, Seoul, Republic of Korea.

City of Cape Town web site: www.capetown.gov.za. Accessed January 2015.

City of Cape Town Strategic Development Information and GIS Department (SDI&GIS). (2013). Data sheets: 2011 census data. Available on www.capetown.org.za. Accessed January 2015

De Wet, Mail and guardian online, 23 March 2012.

Diani, M. (2000). Social movement networks virtual and real. *Information, Communication and Society, 3*(3), 386–401.

Ferebee, S. S., & James W. D. (2009). Factors that persuade continued use of Facebook among new members. In *Proceedings of the 4th International Conference on Persuasive Technology*. ACM.

Gibson, R. K., & McAllister, I. (2011). Do online election campaigns win votes? The 2007 Australian "YouTube" election. *Political Communication*, 28(2)

Hamel, P., Lustiger-Thaler, H., & Mayer, M. (Eds.). (2003). Urban movements in a globalising world. London: Routledge.

Heeks, R. (2001). Understanding e-Governance for development, i-Government Working Paper Series, Paper No. 11.

Howard, P. (2011). *The digital origins of dictatorship and democracy. Information technology and political Islam.* Oxford: Oxford University Press.
Hussain, M. M., & Howard, P. (2012). Democracy's fourth wave? Information technology and the fuzzy causes of the Arab spring, unpublished paper presented to the meeting of the International Studies Ass, San Diego, April 1–4.
Isaías, P., Miranda, P., & Pífano, S. (2009). Critical success factors for Web 2.0—A reference framework. Heidelberg: Springer.
Jung, J.-Y., Qiu, J. L., & Kim, Y.-C. (2001). Internet connectedness and inequality: Beyond the "divide". *Communication Research.*
Latchem, C., & Walker, D. (2001). Telecentres: Case studies and key issues. Commonwealth of Learning, Vancouver.
Maldonado, E. A., Nicolai A. P., & van Gorp, A. F. (2006). *ICT policies as a means to inhibit social exclusion: the South African case. Social inclusion: Societal and organizational implications for information systems* (pp. 137–150). USA: Springer.
Marichal, J. (2012). *Facebook democracy: The architecture of disclosure and the threat to public life.* USA: California Lutheran University.
Minkley, G. (1998). Corpses behind screens: Native space in the city. In H. Judin & I. Vladislavic (Eds.), *Blank architecture, apartheid and after.* Rotterdam: NAi Publishers.
Monge, P. R., & Contractor, N. S. (2003). *Theories of communication networks.* Oxford: Oxford University Press.
Moodley S, (2006). *The promise of E-development? A critical assessment of the state ICT for poverty reduction discourse in South Africa.* University of Durban Westville Press.
OECD Publishing. (2008). Growing Unequal? Income Distribution and Poverty in OECD Countries, October 2008.
O'Reilly, T. (2005). What is Web 2.0. Article on tim.oreilly.com. Retrieved March 2008 from http ://www.oreillynet.com/pub/a/oreilly/tim/news/2005/09/30/whatisWeb20.html
Parker, D., & Song, M. (2006). New ethnicities online: reflexive racialisation and the internet. *The Sociological Review* 54(3).
Putnam, R. D. (2002). *Democratic phoenix: Political activism worldwide.* Cambridge, UK: Cambridge University Press.
Rahman, H. (Ed.). (2005). *Empowering marginal communities with information networking.* IGI Global.
Rheingold, H. (2002). *Smart mobs: The next social revolution.* Cambridge, MA: Perseus.
Robins, S. (2014). Poo wars as matter out of place: Toilets for Africa in Cape Town. Respond to this article at http://www.therai.org.uk/at/debate. *Anthropology Today, 30*(1), 1–3.
Singh, S. (2010). The South African 'information Society' 1994–2008: Problems with policy, legislation, rhetoric and implementation. *Journal of Southern African Studies, 36*(1).
Social Justice Coalition. (2014). About us (online). Available on http://www.sjc.org.za/about-us. Accessed 2 December 2014.
Social Justice Coalition. (2014). *Our toilets are dirty. Report of the social audit into the Janitorial Service for communal flush toilets in Khayelitsha,* Cape Town: 14–19 July 2014. www.sjc.org.za. Accessed December 2014.
Statistics South Africa (StatsSA). web site: www.statssa.gov.za. Accessed January 2015.
StatsSA (Statistics South Africa). (2011). City of Cape Town municipal profile (online). Available from http://www.statssa.gov.za/?page_id=1021&id=city-of-cape-town-municipalitycityof. Accessed 24 January 2015.
Syed, J., & Ali, A. J. (2010). Principles of employment relations in Islam: A normative view. *Employee Relations, 32*(5), 454–469.
Thompson, D., & Gutmann, A. (2004). *Why deliberative democracy?.* Princeton, NJ: Princeton University Press.
UNHabitat, & ECA. (2008). *The state of African cities 2008: A framework for addressing urban challenges in Africa.* Nairobi: UNHabitat.

Van Dijk, J. (2000). Models of democracy and concepts of communication. In K. Hacker & J. van Dijk (Eds.), *Digital democracy. Issues of theory and practice*. London: Sage.
Vignollet, & Baum, S. (Eds.). (2007). *Creative urban regions: Harnessing urban technologies to support knowledge city initiatives*. Hershey, PA: IGI Global.
Weare, Christopher. (2002). The internet and democracy: The causal links between technology and politics. *International Journal of Public Administration, 25*, 659–692.
Wellman, B. (1996). *For a social network analysis of computer networks: a sociological perspective on collaborative work and virtual community*. Paper presented at the 1996 ACM SIGCPR/SIGMIS Conference on Computer Personnel Research, Denver, CO.
World Wide Worx. (2012). *Internet Report Statistics*, South Africa.

Opening up: Towards a Sociable Smart City

Ingrid Mulder

Abstract Release of public sector information (PSI) through co-creation not only resulted in a policy to opening up public sector data. It can be concluded that using a research-through-design approach, connecting crucial partners can create a sustainable infrastructure to opening up data and fosters further social innovation as well. Experiences with co-creative partnerships show how to deal with thresholds and lower them. The open(ed) mindset clearly has impacted the local municipality, in terms of being open to empowering, open to share and being open to change. Not only do the co-created concepts serve as boundary objects, animating public servants to free up more PSI for re-use and giving them potential fuel for other service design applications, but also the collaboration itself enhanced by design interventions seem to be promising in transforming towards a more transparent and sociable smart city. Differently put, the open data movement in our city is not seen as an end unto itself, but as a means to co-create meaningful applications that enrich people's lives, a new strand of city making: people making and co-designing the city.

Keywords Public sector information · Research-through-design · Open data · Co-design · Smart city · Social city · Open collaboration

1 Introduction

In his New Year's speech, Ahmed Aboutaleb, the Mayor of Rotterdam, said that the crisis made Rotterdam strong (Binnenlands Bestuur 2013). The city had faced substantial financial setbacks, though the international crisis has not discouraged city's initiatives, but rather made Rotterdam strong.

I. Mulder (✉)
Delft University of Technology, Delft, The Netherlands
e-mail: mulderi@acm.org

M. Foth et al. (eds.), *Citizen's Right to the Digital City*,
DOI 10.1007/978-981-287-919-6_9

The Rotterdam. "While across the country the construction sector faces difficult times, in our city still structures of size arise. The Rotterdam, the building by architect Rem Koolhaas on the Wilhelmina Pier is national news sensation. The building is a city within a city, a motor for Rotterdam", said the mayor. He also referred to the Rotterdam Central Station, whose major renovation is almost finished.

Facelift. "The city centre has been given a facelift by numerous attractive projects, such as 'the Meent and Nieuwe Binnenweg', two beautifully restored shopping streets, with the allure of the past and the facilities of the future". Despite the crisis, the port of Rotterdam still shows growth. "The port grows to the epicentre of the world in the field of oil storage and oil trade".

Trust and tolerance. However, not only the construction and port sector demonstrate the success of a city. It is the people that make the city, the Mayor argues. "Together we make the future of Rotterdam. And we do that with tools everyone has: with confidence and tolerance".

Air Singel. As an example, he mentioned the Air Singel, a bridge realized by citizens through crowdfunding and awarded best city initiative in 2012. He also referred to those neighbourhoods where residents have made arrangements to make their living environment safer and cleaner. "Or look to those who help their new neighbours learning the Dutch language".

Peace in the city. These are the things that make life meaningful. These small things help. They lead to peace in the city, says the mayor. "Peace is not an abstract concept, it is a verb. Peace cannot be achieved with large conventions, but starts in dealing with people in the neighbourhood, at work or in the city. It starts with something small, something beautiful. It begins with trust".

The speech of the Mayor nicely illustrates a social city, a city open to empower its citizens in order to drive social change. Social innovation is crucial to every city, though driving social change seems to be even more important for Rotterdam, taken into account the extreme situation in the southern part of Rotterdam. "On the South there is an accumulation of arrears in housing quality, work and income, training and education. Southern districts have a relatively young population, with many children who grow up in a deprived environment. Their parents are poorly educated, have little to spend and the environment gives little incentive to move forward. It can be said that the Southern part lacks a climate of learning and working" (National Program Rotterdam Zuid 2012). In a recent meeting, the director of the National Program for Rotterdam South clearly stated that doing the same in a better and more efficient way is not enough. Differently put, the current situation in Rotterdam requires radical innovation and social change. There are severe problems with youth unemployment, air quality, mobility and education. It calls for "reshaping society in the direction of a more participative arena where people are empowered, learning is central which make policies more effective" (Bureau of European Policy Advisors 2010).

Empowerment is crucial in order to drive such a social change; without willingness and personal commitment of citizens, these challenges cannot be met. Hence, citizens are at the heart of our cities and need to be at the heart of change as well. Although we embrace a "bottom-up" approach, we are aware that civic self-organisation alone is not enough; "in order to succeed at scale, grass-roots innovation and localism need support from a new environment of policy, finance, infrastructure and technology" (Robinson 2014). The current work, therefore, addresses the interplay between top-down policy and planning and bottom-up participatory innovation in opening up public sector information (PSI). In doing so, we envisage a *sociable* smart city that enables transforming society into a more participative domain where participatory innovation takes place. A sociable smart city combines best of two worlds: on the one hand, a social city that is people-centred, values active citizenship and embraces community-driven innovation, and on the other, a smart city that welcomes the possibility of future Internet and related technology-driven innovations, such as open data, Internet of Things, and living labs.

2 Background and Motivation

Living labs are Public-Private-People Partnerships (PPPP). They are European Commission's answer to stimulate open, collaborative and bottom-up models of innovation where citizens are at the centre of the innovation process. Living labs' scope of action has extended from undertaking an isolated activity of user involvement towards covering innovation processes from idea conception to result adoption. In their review, Katzy et al. (2012) motivate that the scholarly body of knowledge is shaping with more publications in peer-reviewed journals and consequently a clearer theoretical underpinning. Whereas much living lab researchers address shortcomings of existing theories and practices, few demonstrate mature methodological approaches and true co-creative practices. A living lab is not only a network of infrastructures and services, but also a network of real people with rich experiences and a way to deal with community-driven innovation. Those experiences are the very things that make a living lab living, and therefore, appropriate methods should capture these social and dynamic aspects (Mulder et al. 2008). The concept of living labs derives from the notion that people are innovators, and envisions environments of open and user-driven innovation. A living lab can be seen as "an experiential environment where users are immersed in a creative social space for designing and experiencing their own future. Policy makers and citizens can use living labs to design, explore, experience, and refine new policies and regulations in real-life scenarios before they are implemented" (McPhee et al. 2012).

The contemporary cityscape is increasingly laced with interactive technologies and sensors. Although many of these technologies are inherently personal in nature, often embedded in an omnipresent smart city infrastructure, we are more connected than ever before, wherever we go. At the same time, things are increasingly connected as well. There is no doubt the digital age is affecting our life, public and private, personal and collective. In other words, cities are increasingly becoming a living lab itself, a playground for innovation and transformation, and a playground for co-production of new public services.

Next to that, open data is seen as a promising concept to boost the co-production of new public services. It helps to increase citizen involvement as well as transparency and to improve the decision-making process in general (Bertot et al. 2010; Kroes 2011; Obama 2009). This increased call for data released by governmental agencies, however, is often motivated with a provocative tone of voice. Citizen's right to the digital city seems to go hand in hand with citizens' distrust in government and wish to take over control. In the current work, we aim at a sociable smart city that embraces participation, autonomy and social innovation instead to explore the benefits that open data can have. We use living labbing as an approach to enable co-creative practices in Rotterdam, which help citizens shaping their own surroundings, making and co-designing the city of Rotterdam (Mulder 2012). Or differently put, using the data-rich urban scale as a space for experimenting aiming to turn the city into a participatory domain. In the remainder, we elaborate upon our experiences to open PSI through co-creative partnerships.

3 Approach

On 21 April 2010, the Rotterdam Open Data initiative was launched by innovators from the Rotterdam University of Applied Sciences, Rotterdam-based businesses and creative industry, and the Municipality of Rotterdam to collaboratively explore the value of open data for Rotterdam (Scienceguide 2010). These initiators shared an eagerness to learn and interest to explore new technologies, as well as a shared passion for and will to contribute to their city. Time has come for collective action rather than words. The manifesto supporting the Rotterdam Open Data movement's call to action stated:

> "Because we are convinced that it contributes to the *freedom* of the Rotterdammers, to help them make better, informed decisions, because it *strengthens the bond* they have with their city, and because it enables them to *co-create the city* they live in" (Gommans et al. 2010).

Although it is promising, opening up public sector data is not straightforward. There are many thresholds (see Conradie and Choenni 2012) that require changes in governmental structures, policies and organizations. In their elaborate literature review, Fernandez and Rainey (2006) mentioned various factors that can result in chance in public sector organizations, e.g. the need for internal support, support from top management external support, resource provision and the institutionalization of change. Interestingly, Rotterdam Open Data was born as a PPPP. The manifesto and initiators explicitly address the citizens' voice, which is vital for engaging stakeholders from public sector, industry, education and research as well as citizens in a shared process of knowledge production in which they collaboratively envision desired future cities, which we coined co-creative partnerships

(Mulder 2014). Similarly to co-creative partnerships and the PPPP, Carayannis and Campbell (2009) describe how the often-referred triple helix (university–government–industry) can be extended with a fourth helix to a quadruple helix, which explains that with this fourth helix, knowledge of culture, values and life styles, multi-culturalism, creativity, and media are brought into the process. The fourth helix represents and warrants the human scale required for a sociable smart city.

In our search for a co-creative space for experimenting, a close partnership with education was set-up, allowing for data to be released in a safe (and controllable) environment, where local government can experiment with data release. A close partnership with education also allows data to be released in a more goal-driven way, helping to illustrate data release benefits through prototypes and concepts with quicker results than simply publishing data online. The co-created prototypes are also used to generate explicit and tacit knowledge on opening up PSI, which is typical for a research-through-design approach. The Alderman responsible for Employment, Education, Innovation and Participation and the Chief Information Officer (CIO) both explicitly supported the bottom-up movement, which allowed for mixing top-down and bottom-up dynamics in the current co-creative partnerships. The corresponding approach is also referred to as 'living labbing the Rotterdam way' (Mulder 2012, 2014).

3.1 Citylab Rotterdam

Stadslab Rotterdam is the co-creative space that connects the Rotterdam Open Data partners. Stadslab is a creative hot spot open to citizens enabling making and prototyping, a space for participation and co-creation. The lab has an official Fablab motivated from educational needs to prepare students for new trends in digital fabrication and interactive prototyping, which has been extended with an Internet of Things-laboratory with a strong emphasis on electronic and sensor devices and an Applab in which meaningful applications are designed using (Rotterdam) Open Data. Stadslab Rotterdam is equipped with laser cutters, 3D printers, a vinyl cutter, electronic workbenches and micro-controlling and programming tools. An interdisciplinary set-up in combination with a solidly integrated peer learning and DIY (do-it-yourself) approach empowers FabLab users to "make almost anything". Stadslab Rotterdam offers capabilities to everybody, which have until recently been reserved to a few professionals only. The lab provides an infrastructure for experimenting with sensor technology (Sensorlab), digital manufacturing (Fablab) and open data (Applab). In order to fully exploit the lab in our living labbing practices, students are prepared to use the machines and materials of the lab on their own, and to create and use (physical) prototypes in a safe and confident way. Next to an elective course, students can work in the lab as steward. A team of stewards from various educational backgrounds supports experts, less advanced users, students or visitors of the lab in their making activities (see Fig. 1 and Mostert-van der Sar et al. 2013 for more detail).

Fig. 1 The empty space for co-creation (*left*), and the active making and prototyping activities (*middle* and *right*)

3.2 Open Data

PSI becomes open data when released into the public domain. Consequently, it is deemed suitable for re-use by citizens and available for the creative industry to build on and create new services. While the opportunities of opening up PSI are often referred to and supported by strategic mandates, the release of PSI by local governments is not without difficulties. Hence, many smart city initiatives start top-down, with strategic alliances in order to improve the public services and call for citizen participation. The Rotterdam Open Data approach distinguishes itself by doing, just in keeping with the down-to-earth Rotterdam attitude. Support from the alderman as well as the CIO appeared to be key. It provided the necessary room to experiment, before defining policies.

Since open data are not limited to the government as authority and the citizen as user, all public as well as private sector living lab stakeholders participate together with local government, citizens, creative industry and academia (Dalsgaard 2010; Parker and Heapy 2006). We examined how such participation could be applied to PSI release leading to the co-creation of innovative public services that align with identified citizen needs (see Conradie et al. 2012 for details). Co-creation methods and storytelling were used to identify citizens' needs. These storylines then informed the public service designs, to be created by student teams. These stories were also used as boundary objects (Star and Griesemer 1989) enabling communication between citizens and the participating public civil servants from the city council. Boundary objects are artefacts that allow different groups to work together and share knowledge across boundaries. They are helpful in articulating meaning and address multiple perspectives.

Represented by seven participating city council services, each public service department started the project with a client briefing, in which goals or needs from the departments are presented. Ranging from curbing the social economic health differences in the region to making the core service of a particular council service more accessible to a younger or different target group, the goals acted as starting point for a co-creation design process led by students.

4 Co-creative Practices

Initial co-creative efforts resulted in 36 public service ideas, varying in focus and maturity, though all focused on (re-) using PSI; these ideas were presented during a national open data conference. Participants could vote for the winning concept, after which the alderman responsible for employment, education, innovation, and participation presented the top-five concepts. Some service concepts were taken into development, though the main outcome of the current pilot was to let these applications act as objects illustrating the value of co-creation and the potential of the re-use of open data.

4.1 Open Collaboration

The pilot not only provided many insights, but it also had a larger impact on open data and open innovation in Rotterdam. The active participation and co-creation of multiple partners in the early phases of idea generation managed to put open data on the local policy agenda of the Rotterdam municipality. The board of management of the city council decided to allow the release of the City Development Service's PSI as open data, having currently significant amounts of PSI available in an open data store for experimentation and co-creation of public services in Rotterdam. In addition, the project also introduced the participating creative industry partners to the potential of using and re-using PSI and the important role of the creative industry in that endeavour. Citizens played a role by providing the input for the creation of the prototype applications, which in turn act as concrete examples to illustrate the benefit of the cooperation. By animating public servants to free up more PSI for re-use, potential fuel for other service design applications was created.

Co-creating public services initiated by student teams appeared to be a valuable experimentation place, not in the least for the public servants. Hackathons and service jams also acted as a platform where partners with different strategic backgrounds met and discussed the developed applications. The partnership between academia, the creative industry and the public sector was awarded with additional research funding to further ensure the release of PSI. By ensuring participation of the crucial partners, a sustainable infrastructure has been created to co-create public services and foster further innovation with PSI. Rotterdam Open Data initiative has demonstrated that co-creation can also lead to the development of better public services, with citizens and the private sector contributing data by means of crowdsourcing, and it paves the way for more co-creation through open service development (Louwes 2011).

4.2 Open Data Community

While the momentum for open data was embraced, the open data community continued meeting up during hackathons and service jam to discuss the value of open data and continued co-creating public services using open data. From the many pilot projects, a few were selected to deepen opportunities of data reuse and explore the process of data release.

Rotterdam Unlimited, for example, empowers disabled people to navigate with confidence and pleasure through the city without experiencing any obstacles and thresholds, and informing about available facilities. Using their smartphone, users can benefit from a personalized navigation, but can also report and share any updates to keep the actual information accurate. The co-created public service re-uses raw PSI into a meaningful application and gained commitment from a large audience (see Fig. 2).

With the *LED-sphere,* an extra dimension was given to the open data discussions. The LEDs highlight the density of trees, through linking data of threes to the their gps position. Taking the LED-sphere along during a walk in the city, the LEDs intensity increases when the environment becomes greener. In this way, the LED-sphere visualizes invisible data and clearly contributed to clearly to people's imagination, as well as their empowerment in discussing other opportunities.

Another application called *SunnyApp* shows the free spots to enjoy the sun in the city. The application discusses the use of altitude data and the corresponding complexity for dataspaces. With the *RateThem* app, citizens were encouraged to give data back to the municipality and having a say in the maintenance budgeting. RateThem clearly addresses the citizens' right to challenge and encourages participation.

Ultimately, the goal of the developed applications was not the application in itself. Instead, the focus was on understanding thresholds to data release for the local council and creating an environment for experimentation, where data release can be put to the test. Throughout this process, valuable lessons were learned about opening up PSI.

Fig. 2 Rotterdam Unlimited (*left*) and LED-sphere (*right*)

5 Discussion and Conclusions

In particular, through the close connection with education, we managed to release large amounts of data, initially only for use in education. Having experimented with the data, the sets were subsequently made available on an open data portal, for use by a wider audience. The project has also allowed for the release of data that had been previously sold for profit, as a result of concepts developed around data. A strong connection with higher education can be beneficial for the release of open data and the starting of open data initiatives. Although the approach yielded tangible and valuable results, our co-creative approach is time-intensive collaboration and the concrete results from student projects need a sponsor to turn these in sustainable public services.

In the remainder, we discuss how our approach not only enabled co-creative partnerships to open up data and make value of these, but also that various mindsets have been opened.

5.1 Open Mindset

It can be concluded that using a research-through-design approach, connecting crucial partners has created a sustainable infrastructure to co-create public services that foster further social innovation as well. The established open data policy, having all PSI open *unless*, has clearly a larger impact on the open(ed) mindset and innovation in Rotterdam. Not only do the co-created concepts serve as boundary objects, animating public servants to free up more PSI for re-use and giving them potential fuel for other service design applications, but also the collaboration itself enhanced by design interventions seems to be promising in transforming local government into a transparent government enabling a sociable smart city.

5.2 Open Government, Transformed Government

Local municipality has changed its role from top-down service provider towards a facilitator with their Rotterdammer-centred approach. The Municipality of Rotterdam wants to strengthen collaboration with the citizens and organizations. Co-production of new public services using an integration of top-down and bottom-up approaches has impacted the governance model as well and lead to a transformed government, which values co-production as their mindset, communicated by the council's programme. Having a rich tradition in encouraging citizens' initiatives, the current city programme goes even further. It shows that the baseline is set and that the city government wants to go *full speed ahead* being an enabling government, seeing the city as a living lab. In other words, the governance model has flipped, rather than emphasizing citizen participation; the civil servants are encouraged to participate and attune to the citizens' willingness to create liveable environment; and, moreover, to change policies in case, they hinder the co-production on new public services.

5.3 Open up the Urban Interface

Having a unique mobile and wireless infrastructure, enhanced with public screens and open data, the foundations for a smart city are there. PSI is released and available in the city's open data store. Currently, initiatives are launched to add crowd-sourced citizen-generated data as well. The city infrastructure has become much smarter aiming to facilitate the social fabric as well. However, to benefit from the opportunities offered by the smart city concept, citizens need to be able to participate fully in an Internet-enabled society, through adequate skills and the ability to manage their rights online, such as their privacy.

For this purpose, local government launched the City Initiative aiming to facilitate active citizenship. Citizens are consulted and stimulated in building a beautiful and attractive city. The City Initiative supports people who are proud of our city and in realizing their ideas in making the city more beautiful and fun. This is in keeping with the user-centred design principle that the user is expert of his or her own context (Sanders and Stappers 2008). In 2013, citizens proposed over 120 plans. The City Initiative board made a selection of these 120 plans and has nominated 7 plans. As a next step, citizens make their voices heard and vote on their preferred initiative. The municipality will provide up to €2.5 million to realize the plan(s). The huge amounts of citizen initiatives address the existing social fabric, even though it cannot be expected that all citizens are becoming active citizens. In particular, the sketched context of Rotterdam South calls for serious transformations. Social change is necessary; it can and must be improved at South. There is less lag where better education, more opportunities, better housing, better connection between education and work.

5.4 Open Space

Although Citylab Rotterdam is a prototyping workshop for students in the first place, it is also an interface to the city, an active learning environment for practicing making, co-creation, and participatory design skills, though not only for students, the lab is open to everyone. The fact that the Fablab's doors are said to be open for anyone does not automatically mean everybody will find their way to the lab, especially when people do not have a reason to do so. We have organized various co-creation events, such as hackathons and prototyping workshops mostly related to our open data initiative. Interestingly, not only creative professionals or peer students came to the lab and consequently used the workshop equipment, but also newcomers visited Rotterdam's Citylab. The lab's open identity contributes to an inspiring learning environment. Teachers and researchers from different educational institutes have found the lab to get inspired with new techniques, design and methods for educational purposes. Civil servants and policy makers are visiting the lab as an inspirational zone. Increasingly more citizens find their way to the lab and experiment with new digital fabrication techniques.

For example, primary schools came as part of their "technasium" education aiming to inspire and challenge young people to study science and technology. Similarly, a network organization of businesswomen contacted us for a Technology Safari; mothers together with their daughters participated in a workshop on DIY jewellery.

Trial workshops on 3D printing have been tested with youngsters from a low SES neighbourhood who dropped out the educational system (Pucci and Mulder 2014). In these workshops, they are made familiar with different open-source tools such as (1) Thingiverse, an crowd-sourced collection of open designs created for 3D printing and laser cutting; (2) Tinkercad, an open online 3D modelling tool to create personal designs for 3D printing and laser cutting; (3) Instructables, an online How To and DIY community where people share their inspiring projects and inventions; and (4) the Ultimaker 3D printer, which is an open-source, affordable and easy to assemble 3D printing machine. Results and observations from these workshops are very promising. It shows that people without any prior knowledge can easily participate and finalize their own design in FabLab Rotterdam. With this latter example, we address the potential of young digital natives by co-creating with them a means for their empowerment through a peer-to-peer talent development platform, aiming on the one hand to increase the level of participation of the youth in their local environment and on the other to allow them to self-promote their value through using open technologies. The latter project demonstrated the transformational role of digital techniques: young often-stigmatized adults became active co-creators in the workshops while collaboratively learning new skills. Key to the project's success was the ability of leveraging the students' technological fluency by making digital fabrication accessible and engaging through a participatory bottom-up approach of making and prototyping (Pucci and Mulder 2015).

6 Open Future

Rotterdam Open Data is a movement, a multi-helix co-creative partnership joining forces in promoting the value of opening up data for reuse. It can be concluded that using a research-through-design approach, connecting crucial partners has created a sustainable infrastructure to open PSI in order to co-create public services that foster further social innovation as well. It can be concluded that the open(ed) mindset has a larger impact on innovation in Rotterdam. Not only do the co-created concepts serve as boundary objects, animating public servants to free up more PSI for reuse and giving them potential fuel for other service design applications, but also the collaboration itself enhanced by design interventions has had an important role in transforming local government. Experiences with co-creative partnership show how to deal with and lower thresholds and being open to empowering, open to share and open to change.

Rotterdam Open Data developments feed further discussions within Dutch cities. Since June 2013, the Rotterdam Open Data Store is used in Amsterdam and a few months later in Eindhoven as well. However, it should be noticed that it is not the technological infrastructure that makes the difference; it is the social fabric, the people that co-create the city that demonstrates the value of being open.

We used the data to co-create things to communicate the value and to increase our understanding eventually leading to co-creating meaningful applications that enrich people's lives. Differently put, the open data movement in our city is not seen as an end unto itself, but as a means to co-create our future. The City Initiative and the Right to Challenge manifest this new way of co-creation. This makes the city of Rotterdam a welcome playground for participatory city making. Using ongoing Rotterdam-based projects as design interventions, rich insights can be gained at the boundaries of disciplines, which provides the opportunity to scale and self-sustain the findings. Prerequisites for a sociable smart city are in place; a city as a true living lab enables co-creative practices and has citizens shaping their own surrounding; thus, citizens making and co-designing the city.

Acknowledgments Special thanks to all partners in the Rotterdam Open Data initiative and colleagues of the Rotterdam University of Applied Sciences (researchers of Creating 010, teachers and students of the School of Communication, Media and Information Technology who were involved in the Rotterdam Open Data programme). The work presented has been partly funded by Rotterdam Media Commission and SIA RAAK Publiek. Part of this work has been presented in the Digital Cities 8 workshop in Munich and shared among the workshop participants.

References

Bertot, J. C., Jaeger, P. T., & Grimes, J. M. (2010). Using ICTs to create a culture of transparency: E-government and social media as openness and anti-corruption tools for societies. *Government Information Quarterly, 27*(3), 264–271.

Binnenlands Bestuur (2013). Crisis maakt Rotterdam sterk (Crisis makes Rotterdam strong). Retrieved January 8 from http://www.binnenlandsbestuur.nl/bestuur-en-organisatie/nieuws/crisis-maakt-rotterdam-sterk.8667230.lynkx.

Bureau of European Policy Advisors. (2010). *Empowering people, driving change: Social innovation in the European Union*. Luxembourg: Publications Office of the European Union.

Carayannis, E. G., & Campbell, D. F. (2009). 'Mode 3' and 'Quadruple Helix': toward a 21st century fractal innovation ecosystem. *International Journal of Technology Management, 46*(3), 201–234.

Conradie, P. & Choenni, S. (2012). Exploring process barriers to release public sector information in local government. In *Proceedings of ICEGOV 2012*. Albany, New York, pp. 5–13.

Conradie, P., Mulder, I., & Choenni, S. (2012). Rotterdam open data: Exploring the release of public sector information through co-creation (pp. 187–196). In *Proceedings of ICE 2012*, Munich, Germany. Available online: http://dx.doi.org/10.1109/ICE.2012.6297651.

Dalsgaard, P. (2010). Challenges of participation in large-scale public projects. In *Proceedings Participatory Design Conference 2010*, New York: ACM Press, pp. 21–30.

Fernandez, S., & Rainey, H. G. (2006). Managing successful organizational change in the public sector. *Public Administration Review*, 168–176.

Gommans, L., Mulder, I., & van Waart, P. (2010). Rotterdam open data manifesto. Meaningful Design in the Connected City, Rotterdam University of Applied Sciences. http://docent.cmd.hro.nl/waarp/BOidS_2eworkshop_flyer_ROD_01.pdf

Katzy, B.R., Kuwar, K.S. & Thoben, K-D. (2012). Editorial: A living lab research agenda. *International J. Product Development, 17*(1/2), 1–7.

Kroes, N. (2011). Public data for all—opening up Europe's public sector: 2011. http://blogs.ec.europa.eu/neelie-kroes/public-datafor-all-–-opening-up-europes-public-sector/.

Louwes, K. (2011). Rotterdam open datalab van start met 36 Rotterdam-apps in wording. Available online: http://korrielouwes.nl/2011/07/07/rotterdam-open-datalab-van-start-met-36-rotterdamapps-in-wording/.

McPhee C, Westerlund M, & Leminen S 2012, 'Editorial: Living Labs', *Technology Innovation Management Review*, September 2012, pp. 3–5.

Mostert-van der Sar, M., Mulder, I., Remijn, L., & Troxler, P. (2013). FabLabs in Design Education (pp. 629–634). In *Proceedings of E&PDE 2013, International conference on engineering and product design education*, 5-6 September 2013. Dublin: Dublin Institute of Technology.

Mulder, I. (2012). Living Labbing the Rotterdam Way: Co-Creation as an Enabler for Urban Innovation. *Technology Innovation Management Review*, 39–43.

Mulder, I. (2014). Sociable smart cities: Rethinking our future through co-creative partnerships. In N. Streitz and P. Markopoulos (Eds.). *Proceedings of Distributed, Ambient, and Pervasive Interactions 2014 (DAPI 2014), LNCS 8530*, pp. 566–574. Switzerland: Springer International Publishing. http://rd.springer.com/chapter/10.1007/978-3-319-07788-8_52.

Mulder, I., Velthausz, D., & Kriens, M. (2008). Living methodologies: Understanding the dynamics of innovation. In J. Schumacher & V.-P. Niitano (Eds.), *European living labs—a new approach for human centric regional innovation* (pp. 31–38). Berlin: Wissenschaftlicher.

National Program Rotterdam Zuid, Uitvoeringsplan 2012–2014. Available online: http://www.rotterdam.nl/BSD/Document/Perskamer/RapportNPRZ%20dd%204%20juli%202012.pdf.

Obama, B. (2009). Transparency and open government/The White House. http://www.whitehouse.gov/the_press_office/Transparency_and_Open_Government.

Parker, S., & Heapy, J. (2006). *Journey of the interface*. London: Demos.

Pucci, E. L., & Mulder, I. (2014). IK BEN STER(K). Empowering young adults through a peer-to-peer talent development platform. In *Proceedings of FabLearn Europe 2014*. Aarhus, Denmark. Available online: http://fablearn.eu/wp-content/uploads/fablearn14_submission_36.pdf.

Pucci, E. L., & Mulder, I. (2015). Star(t) to shine: Unlocking hidden talents through sharing and making. In N. Streitz and P. Markopoulos (Eds.). *Proceedings of Distributed, Ambient, and Pervasive Interactions 2015 (DAPI 2015), LNCS 9189.* (pp. 85-96). Springer. http://link.springer.com/chapter/10.1007/978-3-319-20804-6_8.

Robinson, R. (2014, April 8). No-one wants top-down, technology-driven cities. They'd be dumb, not smart. The Urban Technologist. Available online: http://theurbantechnologist.com/2014/04/08/no-one-wants-top-down-technology-driven-cities-theyd-be-dumb-not-smart/.

Sanders, E. B.-N., & Stappers, P. J. (2008). Co-creation and the new landscapes of design. *CoDesign, 4*(1), 5–18.

Scienceguide (2010, April 14). Rotterdam kiest voor Open Data. Retrieved from http://www.scienceguide.nl/201004/rotterdam-kiest-voor-open-data.aspx.

Star, S. L., & Griesemer, J. R. (1989). Institutional ecology, 'translations' and boundary objects: Amateurs and professionals in Berkeley's Museum of Vertebrate Zoology, 1907-39. *Social Studies of Science, 19*, 387–420.

Part III
City Making

Hybrid Participatory Budgeting: Local Democratic Practices in the Digital Era

Stefano Stortone and Fiorella De Cindio

Abstract Participatory budgeting (PB) is an innovative democratic practice that consists of giving citizens in a local community the power to identify spending priorities, put forth and develop concrete proposals, finalize them into feasible projects, and select which projects are worth financing and implementing. It gained attention for its capacity to include people in policymaking at the urban level. Begun as a face-to-face participatory process, PB today is shaped by ICT that combines online and offline participation venues. This chapter presents this new wave of "hybrid" PB and analyzes citizen participation data from nine recent Italian case studies in six municipalities that have benefited from a Web platform specifically developed for PB processes.

Keywords Participatory budgeting · E-democracy · Online deliberation · E-participation · Deliberative democracy · Smart citizens · Smart city

1 Introduction

Information and communication technology (ICT) increases opportunities for citizens to craft their daily urban life (see among the others Aurigi and De Cindio 2008; Foth and Satchell 2011). It helps people save time and money in obtaining

S. Stortone (✉) · F. De Cindio
Università Di Milano, Via Comelico 39/41, 20135 Milan, Italy
e-mail: stefano.stortone@unimi.it

F. De Cindio
e-mail: fiorella.decindio@unimi.it

M. Foth et al. (eds.), *Citizen's Right to the Digital City*,
DOI 10.1007/978-981-287-919-6_10

and sharing information or services and in coordinating their activity toward a common goal. What was originally bizarre hitchhiking has now become "smart," widespread carpooling, while everyone today enjoys passing the hat for pet (small or large) dreams through crowdfunding. Open data from public agencies are ultimately a gold mine for start-ups and their apps. All these changes share a common characteristic: they are relational improvements in the private, economic sphere toward a more democratic and public form. Time, money, and knowledge are like democratic votes, through which people contribute to shaping and making things happen, building individual desires into collective projects. Intermediation is reduced to a minimum, sometimes just to an essential software platform.

Some of these issues also arise at the political level. Public opinion can benefit from more tools and opportunities to affect the political sphere and shape the urban context. Social reporting, open data, citizen journalism, and crowdsourcing are just few of the innovative, empowering practices to date. In some cases, they make cities "smarter" and more participatory. In others, they enable millions of people to take to the streets and trigger revolutions through the use of social networks, whether by the Indignados in Spain, the #Occupy movement in the USA, the V-Days in Italy or the Arab Spring in North African countries (Castells 2012). The Web has even nurtured the emergence of new and innovative political forces (the Pirate Party, M5S, and Podemos, in Germany, Italy, and Spain, respectively) whose success is grounded in strategic use of the Internet.

Such innovations are gradually changing democracy. However, people are still far from having broad impact on policymaking, and policies are still far from being elaborated in a collaborative way. There are two main problems. On the one hand, such innovative ICT solutions are usually top-down products designed more for the purposes of the owners, such as collecting and using user-generated content, than on those of the people—to gain real democratic power (Foth and Brynskov 2016). They are mainly proprietary solutions which cannot be freely (re-)used or customized. On the other hand, political representatives find it difficult—or make little effort—to design effective and appropriate participatory processes, no matter what technology is employed. Even the newest, most technology-oriented political organizations still rely on strong leadership (Castells 2009) and use rather trivial software to involve citizens, as the case of the Beppe Grillo's Five Star Movement in Italy (Bordignon and Ceccarini 2013) as well as Podemos (Iglesias Turrion 2014) in Spain show.

To sum up, despite ICT's great potential, citizen participation is still highly fragmented and limited to individuals' private sphere and to minor public issues. When set up by political actors and political institutions, online and offline participatory processes seem unable to move beyond advisory and weak forms of citizen involvement (Bertone et al. 2015). Whoever the "owner" of the initiative and of the related technology is—the government, a political party, or a company—the user-citizens have to accept their *terms and conditions* and features chosen top-down, according to the owner's (political or commercial) objectives. In any case, people find it difficult to play a powerful and creative role in policymaking. They are still in the back seat and have to fight to get things, because they do not have the capacity and the power to *do it themselves* (Antoniadis and Apostol 2014). Mobilization is a good achievement but always comes later and is mainly about opposing rather than proposing. In any event, even after results have been obtained, the problems of inclusion and participation arise anew. This context may even make a city "smart," but it is still far from making smart citizens.

Participatory budgeting (PB) seems to address some of these issues. PB is a relatively new democratic invention. Unlike other similar initiatives, it structures participation into an articulated, comprehensive process where citizens not only contribute but are directly involved in final decisions. People decide directly how to spend public money and, more generally, make decisions affecting their daily urban lives, backed by strong commitment from government and clear rules. Devised in Brazil at the end of the 1980s as pure, face-to-face participatory practice, PB quickly spread to thousands of cities worldwide due to its positive results in terms of citizen involvement and democratic outcome (Avritzer 2009; Sintomer et al. 2012, 2013; Allegretti et al. 2013). Thanks to the increased use of the Web for political purposes and the current financial crisis, which called for more transparency, openness, and clarity in public policy and budget management, PB has now been brought back to the fore. PB projects are now using digital tools and online spaces to increase information, improve transparency, and remove barriers of space and time, thus gaining legitimacy. However, opinion is divided: the Internet increases participation beyond traditional, face-to-face PB but may reduce deliberation and social relations among citizens, fostering individualistic behavior (Allegretti 2012).

Actually, there is a kind of natural connection between PB and online interaction. In both cases, people participate directly by submitting, debating, supporting ("liking"), and thus aggregating around issues and ideas, rather than through predefined political membership or platforms. Beyond spreading knowledge of PB worldwide, the Internet appears to enable a certain kind of participation that can find an institutional way into the PB process. It is this synergy between PB and the Internet that leads us to suggest PB as a suitable practice to be considered in the debate on *citizens' rights to the digital city*. This calls for testing two main questions: on the one hand, whether PB makes citizens smarter, empowering them to participate in policymaking; on the other, whether PB fosters online deliberation and nurtures online participatory tools, given its specific methodology. Succinctly put, it is useful to evaluate the impact of PB on citizen participation as a whole, both offline and online.

This chapter begins answering these questions by describing and analyzing some recent Italian case studies of this innovative democratic process. These initiatives are all hybrid cases of PB, combining online and offline venues for participation. They use the same Web platform, developed specifically for PB, but do not share the same PB process. As we shall see, these are design choices which affect citizens' participation. The chapter is organized as follows. Section 2 reviews how PB works and how it has evolved from face-to-face to hybrid processes. Section 3 presents the design guidelines and features of the software platform, "BiPart."

Section 4 illustrates the nine PB initiatives, and Sect. 5 summarizes their outcomes comparatively. The conclusion discusses these outcomes with respect to the impact of PB on citizen participation, especially when ICT-enabled.

2 Participatory Budgeting: Offline and Online Practices

PB is a democratic practice that directly involves people in making decisions on public expenditures. It began in 1989 in Porto Alegre, Brazil, after the fall of the military regime, as a result of the pressure exerted by emerging social movements and by the election of the Workers' Party to city government. Its aim was to overcome the limits of liberal democracy by guaranteeing redistribution, social justice, inclusion, and social cohesion, as well as greater public accountability. PB drew the attention of international institutions such as the World Bank (Shah 2007) and UN-Habitat, who named it one of the world's best practices for local governance. Since then, PB has quickly gained broad consensus (Dias 2014), spreading to thousands of cities, worldwide, and growing in quality. Just to mention the largest cities outside Brazil, PB has been implemented in Rosario, Argentina, Paris, France, Cologne and Berlin, Germany, Turin, Milan, and Monza, Italy, Lisbon and Cascais, Portugal, Cordoba, Malaga, and Seville, Spain, and New York, Chicago, and San Francisco in the USA.

PB can be a relatively simple process. It may also evolve into a more complex structure over time. Essentially, it is a learning process that consists of giving citizens in a local community the power, every year, to identify spending priorities, make and develop concrete proposals, finalize them into feasible projects, and select which ones are worth financing and implementing. Citizens carry out this process directly or by appointing fellow citizens to represent them and their priorities, especially when interacting with local authorities and with other stakeholders. As we will see, there are many kinds of PB worldwide, but ultimately, PB consists of three main phases in which citizens (Fig. 1):

1. prioritize public issues, make and preselect proposals, and appoint delegates, if needed, to represent and carry them out;
2. along with the municipal officials co-design collected proposals into feasible projects whose cost has been established;
3. hold a final vote on the projects to choose what ones to implement.

The PB process is usually preceded by a preliminary public debate on the city's main issues, possible solutions, and the available resources. Once voted on (phase 3), winning projects are then implemented and their implementation can be monitored by the community at large. These two further phases are also depicted in Fig. 1 but do not belong to the PB process, strictly speaking.

Participation is often extended to all city dwellers age 16 and older. PB usually involves only a portion of the overall municipal budget, primarily investment in public works (e.g., parks, street maintenance) and, sometimes, public services

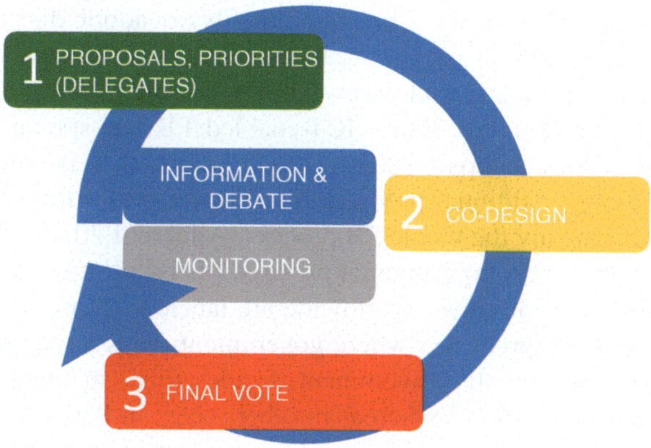

Fig. 1 The participatory budgeting cycle

(e.g., welfare or cultural initiatives). But it may also extend to other policy areas (taxes, etc.). Generally speaking, it is a wide-ranging innovative model for political decision making that allows needs and priorities, and their representatives from the community, to emerge and to connect to one another (Stortone 2010). In the most advanced PB, the citizenry also contributes to setting the rules for the process, explicitly established in a statute.

Despite this institutional democratic nature, formal power is retained by traditional representative bodies—the mayor, the executive body, and the city council—which hold democratic legitimacy by virtue of election. They commit themselves to carrying the participatory process out and to implementing the projects citizens propose and select. Although important, the informal character of this "participatory contract" (De Cindio 2012) still characterizes PB as an option (or an opportunity), rather than a political right, whose functioning is firmly in the hand of the municipality. This affects the degree of people's involvement: the greater the haziness, the less incentive there is to participate. Participation is also penalized when PB relies only on demanding face-to-face meetings, which are usually attended by the most active citizens and members of civil society organizations. In its most successful incarnation, in Porto Alegre, PB was attended by a small part of the population, one or two percent. PB has always struggled to reach significant levels of empowerment and to attain enough legitimacy to impose itself as an established bottom-up practice in local governance. This is one of the reasons why it is often short-lived in many cities, stopping as quickly as it starts.

In order to overcome these shortcomings, PB is now assisted by ICT (Matheus et al. 2010; Allegretti 2012; Sampaio 2014; Sampaio and Peixoto 2014). Almost every PB project today has its own Web site, often an information space, as it happens in Porto Alegre (portoalegre.rs.gov.br/op/). Some, as in Chicago, use online forms to gather citizens' proposals (pbchicago.org/49th-ward.html). Others have interactive maps for citizens to report and propose, like New York City (pbnyc.org/idea) and recently Paris (budgetparticipatif.paris.fr), as well as forums and Facebook pages to debate online. Belo Horizonte (more than 2 million inhabitants) in Brazil was the first municipality to adopt digital PB (opdigital.pbh.gov.br). Since 2006, every two years, hundreds of thousands of citizens have voted on a list of projects, exclusively via the Internet, mobile phones, and electronic voting booths (Peixoto 2009). ICT-enabled PB has spread exponentially worldwide, as shown in Sampaio (2014). Nevertheless, ICT is still mainly used to sustain prevailing traditional procedures, typically performing the final vote. It does not yet accompany the whole PB process (Allegretti 2012).

Being an experimental process, PB is still a work in progress. Therefore, many different initiatives worldwide are labeled—more or less properly—PB. There are *weak* PB processes where government plays a major role: in some cases, citizens propose but the government decides (e.g., in many German cities; Scherer and Wimmer 2012),[1] whereas in others (e.g., in Paris, in 2014), the government establishes what projects citizens can vote on. *Stronger* PB processes are *deliberative*, in the sense that citizens have the right to propose (phase 1), design (phase 2), and vote (phase 3). Among deliberative PB processes, there are interesting differences that depend mainly on the mode and degree of civic engagement in making proposal: (a) some select a small sample of citizens who are driven into a demanding series of face-to-face meetings (e.g., Turin) or organize ad hoc public meetings (e.g., Cascais); (b) others offer everyone the chance to make proposals, especially through Internet, and then to select the most popular among those that are feasible (e.g., Paris in 2015) or work together with the proposers of those most widely agreed upon (e.g., Monza).

These two different approaches to PB can be related to two main theories of deliberative democracy, as identified by Mansbridge et al. (2012). The first approach, which we will call type A, is influenced by what they call *ideal proceduralism*: it looks at a more "shielded" process to reduce the role of organized groups and to increase the deliberative capacity of this "minipublic," who must be well informed to be in a position that favors making the best decision acceptable to all (see Bobbio 2010). The second, type B, draws inspiration from their own *systemic approach*: it aims to create a well-designed, democratic, deliberative system by maximizing the chances for everybody to be part of the process at any phase, in order to increase the legitimacy of the decision, and by strengthening social relationships and people's knowledge and skills.

There is already a mutual contamination between these two approaches. For example, almost all PB processes today have face-to-face meetings to strengthen social ties and converge toward shared projects, as well as a final (online and/or offline) citywide vote to legitimate the winning projects. The choice of participation modality—namely type A or type B or something in between, the choice of software—what tool and which features to include at configuration time, and the mix between online and offline venues determine the design of each PB instance. This affects online and offline participation. It influences the cost of carrying out

[1]Recent research (Allegretti and Dias 2015) shows that these forms of PB are more likely to fail.

the PB process and its scalability. For instance, providing mobile polling stations increases participation but also organizational expenses.

Italy is a good field to observe the interplay among these factors (Allegretti and Stortone 2014). PB arrived in Italy in the early 2000s, mainly promoted by left-coalition local governments. Before long, hundreds of municipalities undertook it. These early initiatives achieved little significant outcome, partly because they mainly adopted the weak form of advisory PB. As a result, after an initial boom, PB rapidly disappeared. Today, there is a new wave of PB in the country, due to the interplay of a couple of key factors. One is the global economic crisis and its impact on quality of life. Another is the rise of social movements that call for radical change in economics and politics, which have exploited the Internet as a mobilization platform (Castells 2012). In this scenario, PB is seen as a way to answer citizens' demand to "be part" of it. Two regions, Emilia Romagna and Tuscany, have already adopted measures to promote participatory processes, including PB. At the same time, young activists and elected councilors, typically from the Five Star Movement (M5S), have pushed for the adoption of PB from the bottom, up. The vast number of academic studies on participatory and deliberative democracy has given scientific support to this new wave of PB processes. Whereas PB processes in the early 2000s were managed—often ad-libbed—directly by municipalities, now experts and consulting bodies are being hired to run PB, to assure quality, and even to provide independence from political influence.

The following paragraphs describe and analyze case studies from Canegrate, Cascina, Cernusco Lombardone, Faenza, Turin, and Monza. We chose these cases because they are all deliberative PB processes, representative of the two approaches described so far, and all adopt both online and offline voting. Because they shared a common Web platform, BiPart, we could query online participation data directly.[2] Many of them were run by the Center for the Study of Participatory Democracy (CSDP), a civic association founded by the first author to promote PB, which also designed and developed the BiPart Web platform, which provides us data access. Its characteristics are outlined in the following section.

3 Design Principles and Development Trajectory of BiPart

A simple, low-cost mash-up Web site that was nevertheless easy to use was developed in 2010 for the first PB initiative carried out by the CSDP in the town of Canegrate, Lombardy (12,000 inhabitants). It allowed citizens to be informed (and stay up to date with content from the social networking sites), to upload their proposals, and to vote. However, it did not offer any sort of interaction. The first release of the BiPart software platform (pronounced "be part," after the Italian BP

[2]This was not the case with other significant, Italian PB initiatives, such as in Rho (dirolamia.it/) and Capannori (comune.capannori.lu.it/node/9066).

for *Bilancio Partecipativo*) was then developed in 2012, to support three new PB initiatives: in Cascina, Tuscany, in Cernusco Lombardone, Lombardy, and in the Province of Pesaro and Urbino; all coordinated by the CSDP.

The design idea behind BiPart is: (a) to create a digital "civic space" where people can freely gather, share, debate, and aggregate around issues and proposals, strengthening relationships and self-organization; (b) to provide a more institutional, deliberative space that mirrors the PB process and leads people to make decisions; (c) to assist several parallel PB initiatives, thus keeping costs low while facilitating the sharing of experiences among them and encouraging the creation of a broader community of practice focused on PB; and (d) to allow any community to develop their own PB process, independent of government initiative.

BiPart is a *multi-site platform* that can accompany as many PB processes as needed. Applying the framework introduced in De Cindio (2012), each instance is now equipped with the following features:

- *Community space*: Registered citizens can report problems, publish ideas or more detailed proposals, open local or thematic groups, and collect followers or members around each of these items. They can do it over time. As groupware technology, each of these items includes room for comments, a notification system, and storage for uploading multimedia documents. Items can also be georeferenced, associated with predefined categories, tagged or shared through popular social networks.
- *Deliberative space*: Each community, usually through an intermediary proposing institution, such as the municipality, can set up a PB process according to a well-defined schedule and budget. Platform administrators customize the process accordingly. They select the appropriate features for each phase and activate the deciding actions that allow citizens to support ideas and proposals, to prioritize issues, to appoint delegates, and ultimately to vote on the projects. When PB is underway, the home page allows citizens to follow the process and be part of ongoing activities. Registered citizens who want to perform deciding actions can be required to authenticate by supplying further credentials (such as taxpayer identification number) and a text-message verification code received to their cell phone. People can even vote on-the-fly, without an account. Data can then be verified (automatically or by municipal offices) to validate participation. All these features are active only while a PB process is in progress.
- *Personal space*: Each registered user has a profile to edit, a dashboard of activities with a relationship map (e.g., problems and ideas s/he published, group memberships, etc.), and a timeline that logs activity on the Web site. In order to facilitate cross-community fertilization, people can participate in different instances of BiPart, so this timeline includes all the items and news feeds that each user joins and follows within the platform.

BiPart was developed incrementally, in close coordination with ongoing initiatives and available financial resources, and tested in real-life situations. Table 1 summarizes the key features of BiPart's five versions: 1.0, 2.0, 3.0, 3.1, and 3.2. BiPart 1.0 was developed as old-fashioned, Java-based EDP software that simply

Table 1 BiPart versions and their tools and features

Features/versions	v. 1.0	v. 2.0	v. 3.0	v. 3.1	v. 3.2
Proposals (+support)	yes	yes	yes	yes	yes
Problems	no	no	yes	yes	yes
Ideas (+support)	no	no	no	no	yes
Groups	no	yes	yes	yes	yes
Voting page (projects)	no	yes	yes	yes	yes
SMS authentication	no	no	yes	yes	yes
Calendar	no	no	no	yes	yes
General forum	no	no	no	yes	yes
Notification system	no	no	no	no	yes
Personal dashboard	no	no	no	no	yes
Community timeline	no	no	no	no	yes

let people make or support proposals. It was quickly withdrawn because of limited usability and a new development team brought in. BiPart 2.0 was developed on the Linux, Apache, MySQL, Perl/PHP/Python (LAMP) platform. BiPart 3.0 had a substantially revised user interface and introduced the text-message authentication system. Other features (calendar, forums, notifications, timeline, and so forth) were incrementally added to the current version, 3.2. The software (and links to all associated PB Web sites) can be found at bipart.it. It is available as a service through CSDP.

4 The Case Studies

This section briefly describes the nine PB case studies. They are all deliberative PB and cover the two approaches described so far. They mainly differ on how people's needs and proposals are collected, filtered, and then turned into projects. All our PB cases adopted approval voting, whereby voters can choose more than one alternative, except for Turin, where plurality voting meant voters could choose only one alternative. Moreover, in all these PB initiatives, residents age 16 and above could participate. Table 2 includes other parameters of context (population, city budget, PB budget, organizational costs), including the varying technology (as BiPart versions, cf. Table 1) and who coordinated the PB process. These statistics help contextualize and understand participation outcome, as outlined in Sect. 6.

Canegrate Partecipa!—the Canegrate municipal PB initiative—was carried out in 2010, 2011, and 2013. The PB structure was simple: (1) citizens make proposals and ask fellow citizens to propose the same thing; (2) municipal offices evaluate the feasibility of the most popular proposals and estimate their cost; (3) estimated proposals are put to a vote to determine which will be funded. Proposals (phase 1) and votes (phase 3) were gathered through paper ballots distributed door-to-door and in several public places and then collected in ballot boxes at the same public

Table 2 The PB case studies

PB initiative (city + year)	Population	City budget (in k€)	PB budget (in k€)	% of city budget	PB cost (in k€)	% of PB budget	BiPart (version)	Consultant
Canegrate2010	12,431	11,000	100	0.91	15	15.00	0.1	CSDP
Canegrate2011			150	1.36	17	11.33	0.1	CSDP
Canegrate2013			70	0.64	13	18.57	3.1	CSDP
Cernusco2012	3863	4000	100	2.50	6	6.00	0.1/2.0	CSDP
Cascina2012	44,133	45,000	1000	2.22	55	5.50	1.0/2.0	CSDP
Cascina2013			350	0.78	37	10.57	3.0	Other
Faenza2014	57,973	114,000	100	0.09	20	20.00	3.1	Other
Turin2014 (District 7)	90,000	n.d.	500	n.d.	61	12.20	3.2	Other
Monza2014	120,184	410,000	800	0.20	20	2.50	3.2	CSDP

locations. They were also collected through a basic Google form in the mash-up Web site. In 2012, PB was suspended because of municipal elections. In 2013, the incumbent coalition was re-elected and held a new PB session. The process changed remarkably, with a more sophisticated pilot procedure that BiPart 3.2 enabled: (1) instead of making proposals, citizens assign relevance (from 1 to 10) to each public interest (the environment, urban mobility, etc.) and appoint at most three fellow citizens for their priorities; (2) citizens who garner the most support—representatives for priority issues—become part of a PB council, charged with developing the projects to vote on; (3) citizens vote on the projects.

Cernusco Partecipa! Cernusco Lombardone's municipal PB, was conducted only in 2012. It was set up as in Canegrate 2010 and 2011, but in phase 1 people could also appoint a fellow citizen (delegate) to represent their proposal and to develop the project to vote on. Due to the problems with BiPart 1.0, Cernusco resumed the early mash-up Web site in phase 1, while BiPart 2.0 was already available for the final vote.

Cascina Partecipa! is Cascina's municipal PB initiative. It ran for two years, 2012 and 2013. In 2012, phase 1 was organized as in Cernusco, but the 16 final projects to be voted on were designed and selected (phase 2) through structured offline meetings of delegates for the 20 most widely supported proposals and a representative sample of citizens (80 people). BiPart 1.0 was used for phase 1 (proposals could now be supported by clicking on the original, rather than rewriting as on the mash-up Web site) and BiPart 2.0 was released for the online vote. Citizens could vote on the 16 final projects either at municipal offices for the entire duration of this phase or on about 20 scheduled dates at a mobile polling station. Unlike in phase 1, there was no widespread distribution and collection of ballot papers, since it had proved unsustainable due to the size of the city. In 2013, the PB process was carried out by a new agency more oriented toward the type A. Therefore, the first phase changed considerably. No longer was there competition and preselection of proposals (and delegates). Instead, phase 1 was an informal gathering of citizens' needs, collected both offline, at face-to-face meetings or with paper straw ballots, and online, through BiPart 3.0. A representative sample of 80 citizens was selected to debate those needs and came up with a set of 14 projects to be voted on.[3] People could again vote at municipal offices for the entire duration of the phase or on about 40 scheduled dates at a mobile polling station. Alternatively, they could vote online, by authenticating through the new, stricter authentication procedure based on text-message verification.

OplàFaenza—the Faenza municipal PB initiative—was fairly similar to Canegrate 2010 and 2011. Citizens first made proposals and expressed support. Local government then evaluated the feasibility of the most widely supported proposals, which were voted on. The €100,000 budget was to fund at least five projects. This

[3]In both the editions, the sample of citizens who attended the face-to-face meeting and who ultimately voted on the projects during balloting fell from 80, initially, to 20–30 citizens, in later sessions.

was the first case in which the Web site was the main channel for supporting and voting on proposals: all proposals, even those made in offline deliberative meetings, had to be uploaded because they could only be supported online. Unlike voting, the support action did not require participant authentication. Voting occurred mainly online, with offline voting possible at one dedicated municipal office every business day and at a mobile polling station only on two Saturday mornings.

Apriamo il bilancio! (or "Open the Budget!") is the PB initiative setup in District 7 of the City of Turin, the capital of the Region of Piedmont. As in Cascina2013, phase 1 consisted of preliminary debate over public issues (problems and needs) that were reported through widespread local meetings and on BiPart. A representative sample of 15 citizens—identified on a random basis among those who volunteered—met five times to analyze these issues. They agreed to design three alternative projects to vote on, each claiming the total budget. Voting lasted 10 days online and seven days offline, at one polling place open weekdays.

Partecipa and Decidi, the City of Monza's PB initiative also drew inspiration from Cernusco, in this case to split of phase 1 into two distinct steps. People first submitted proposals, then chose which to support. The idea was to give people the chance to look at and debate all the proposals. The most widely supported proposals were evaluated, estimated, and turned into projects through collaboration between city officials and the proposers. Final projects were then put to a vote. The €800,000 budget was split into €200,000 for services and €600,000 for investment, doled out equally among 10 neighborhoods (€20,000 and €60,000 each). Therefore, each project had to cost no more than one-tenth of the total budget. Ballots were widely distributed and collected only in phase 1, whereas the final vote in each neighborhood was held at one fixed polling place for the whole duration of the phase and one mobile polling station on about seven scheduled dates.

Except for Turin, Faenza, and the two early cases in Canegrate, the other PB initiatives took advantage of explicit rules specified in a statute, the *Carta della Partecipazione*.

5 Comparative Analysis

Table 3 and Fig. 2 show participation results in phases 1 and 3 and the duration of each phase. In addition to features noted in Sect. 4, other factors influenced participation outcome. These included what communication strategy was adopted, the government's actual commitment, and sociopolitical context, to name just few. Space prevents discussing them all in the analysis of the nine cases.

We note, first, that participation in phase 3 (column 7) is usually greater than in phase 1 (column 3). There are two main reasons. First, the municipality tends to make more organizational effort for the final vote, which legitimates the outcome

Table 3 Participation rates in the nine hybrid PB case studies

1	2	3	4	5	6	7	8	9
	Phase 1				Phase 3			
PB initiative (city + year)	Days	Total supporters (% of population)	Online supporters (% of total)	% of population	Days	Total voters (% of population)	Online voters (% of total)	% of population
Canegrate2010	54	426 (3.43 %)	28 (6.57 %)	0.23	42	590 (4.75 %)	115 (19.49 %)	0.93
Canegrate2011	52	656 (5.28 %)	74 (11.28 %)	0.60	48	1232 (9.91 %)	85 (6.97 %)	0.68
Canegrate2013	17	234 (1.88 %)	16 (6.84 %)	0.13	20	351 (2.82 %)	39 (11.11 %)	0.31
Cernusco2012	50	234 (6.06 %)	66 (28.21 %)	1.70	36	550 (14.24 %)	65 (11.82 %)	1.68
Cascina2012	38	2250 (5.10 %)	93 (4.13 %)	0.21	31	1550 (3.51 %)	889 (57.35 %)	2.01
Cascina2013	–	–	–	–	31	1365 (3.09 %)	238 (17.44 %)	0.54
Faenza2014	77	2400 (4.14 %)	2400 (100 %)	4.14	16	1353 (2.33 %)	1219 (90.10 %)	2.10
Turin2014 (District 7)	–	–	–	–	10	1807 (2.01 %)	1712 (94.74 %)	1.90
Monza2014	24	1236 (1.03 %)	738 (59.7 %)	0.61	29	3619 (3.01 %)	1450 (40.07 %)	1.21

Online participation data come from queries on the BiPart platform. Along with offline participation data, figures were certified by the municipal offices for each PB initiative. Specifically, the number of supporters and voters gives the participation rate, per population, in the two phases. Rates of online supporters (column 4) and voters (column 8) refer to total participants during phase 1 (column 3) and phase 3 (column 7), respectively. Rates in columns 5 and 9 refer to online participants (supporters and voters, respectively). Since Cascina2013 and Torino2014 adopted the type A, they yielded no data for support actions

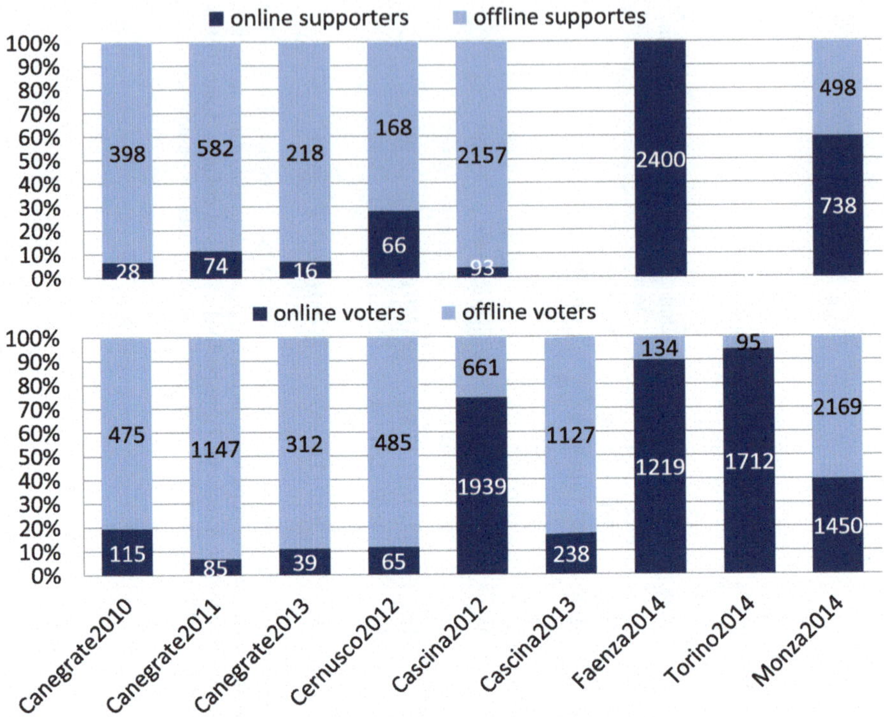

Fig. 2 Participation during phase 1 and phase 3 in the nine case studies

of the whole PB process. Second, voting is a rather simple, familiar activity, directly aimed at deciding which projects will be implemented.

The only two exceptions, Cascina2012 and Faenza2014, can be explained as follows. In Cascina2012, proposing and supporting actions were facilitated by the widespread distribution and collection of paper ballots, which did not take place in phase 3. Moreover, many online votes (roughly one thousand) were thrown out due to irregularities. Taking advantage of the lack of strict authentication (see Sect. 3), some groups had collected proxies from citizens, then enrolled, and voted on their behalf. Had these ballots not been disallowed, there would have been more voters than supporters. In Faenza2014, supporters outnumbered voters two to one. This came about because phase 1 lasted longer (77 days) than phase 3 (16 days) and because there was no authentication, which was subsequently introduced in phase 3 to check voters' identities.

Let us now consider participation during the voting phases. When municipalities invest time and energy in collecting votes at offline venues (e.g., organizing numerous polling stations, as in Cascina and Monza), and/or let people pass out paper ballots or even vote from home (as in Canegrate and Cernusco)—all options spurred by a desire to include more citizens and increase the legitimacy of the decision—offline participation becomes easier and offline turnout proves higher than online turnout. However, it is worth noting that without substantial offline investment, especially in larger cities, the online vote prevails and manages to keep turnout relatively high, obviating a larger drain of voters. This happens even without an equivalent investment in Web campaigning and in less time. Faenza and Turin show the lowest overall participation rate during the voting phase (2.33 % and 2.01 %), but this essentially came from the Web (90.10 % and 94.74 %, respectively) and in very limited time (16 and 10 days).

The analysis of voting behavior in Monza2014 confirms the importance of having a hybrid PB. Figure 3 provides the age distribution of the online and offline voters, and of the total voters compared with the age distribution of the whole electorate (16 years old and above). The shift to the right side of the electorate shows the prevalence of the elderly. We can recognize three patterns. The two segments of the population on the left-hand side are under-represented, but the online participation avoids a more radical loss of the youth. The two central segments—corresponding to adults in their working age, those who can be assumed more interested in the way in which public money are invested—are (slightly) overrepresented in the voters' distribution, and again this has to be ascribed to the online participation. Finally, in the case of two segments to the right-hand side, the rate of the total voters quite well fit the electorate, but, for the oldest people, this is mainly due to the highest rate of offline voters. Actually, almost the 40 % of those who voted offline were above 65 years old, being this, in Italy, the age of retirement. In summary, these data show that the online and offline participation venues are both necessary to involve complementary sectors of the population. Hybrid solutions are still necessary in democratic processes, since the lack of one dimension (online or offline) would cause the exclusion of a segment of the population.

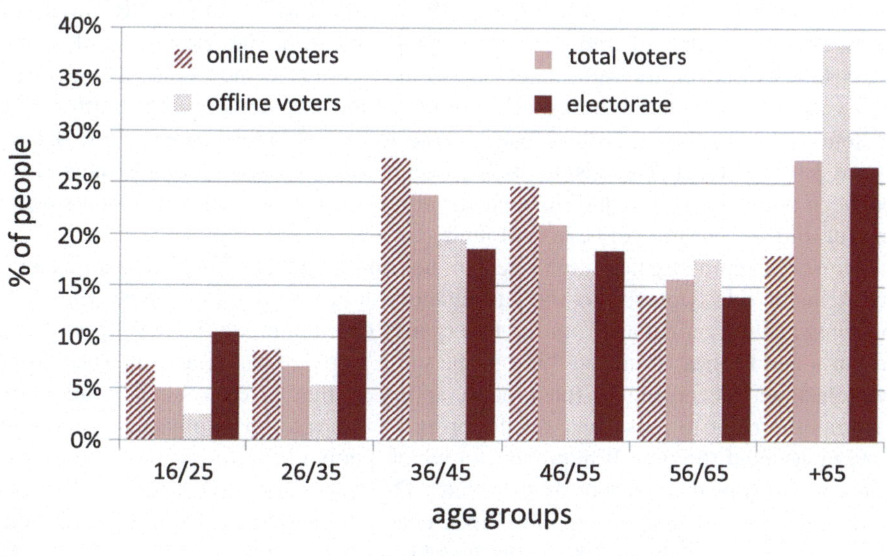

Fig. 3 Age distribution of the voters in the PB of Monza

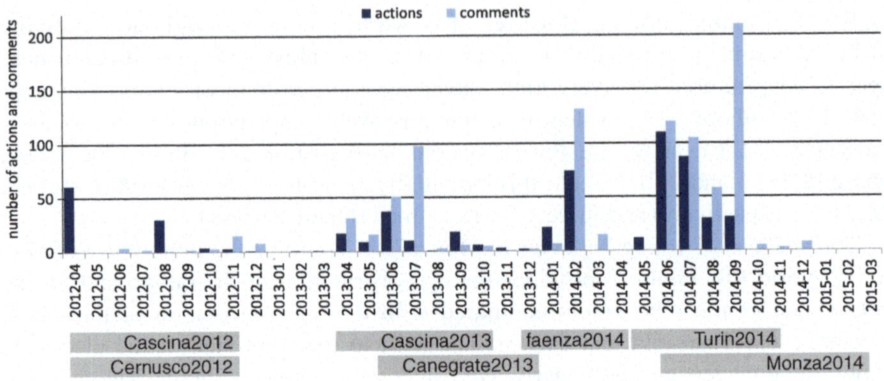

Fig. 4 Online (monthly) activity in BiPart and ongoing PB initiatives

Participation is also affected by software platform usability. Cascina2012 had the lowest rate of online participation in phase 1 (4.13 %), easily attributable to the shortcomings of BiPart 1.0, in contrast to extensive, costly door-to-door distribution of paper ballots. As catches the eye on right-hand side of Fig. 2, the highest online participation rates belong to Faenza (100 % and 90,10 %), Turin (94,74 %) and Monza (59,70 % and 40,01 %), which used the latest version of BiPart and enjoyed a more friendly, although still complex (Anzani 2015), user interface.

Software usability affected the use of BiPart beyond supporting and voting. Figure 4 shows the distribution over time of actions (to create problems, ideas, proposals) and comments. It tracks the instances all together, from phase 3 in Cascina2012 and Cernusco so far.

First, we notice that the peaks correspond to the phases 1 of one PB initiative, when citizens create and comment on relevant content, be they problems, ideas, or proposals. Moreover, the right-hand side of the figure also shows increasing activity easily ascribable to the improved usability of the software (e.g., the notification system). It is worth noting that, again, no action takes place beyond phase 1 and phase 3 (the voting). The design choice of a community space, which stays alive between one PB instance and the next for purposes of monitoring and brainstorming (cf. Fig. 1), has not yet been exploited.

However, online participation rates are not only a matter of the software usability. A major role is played by the participatory approach that is adopted. Figure 5 presents activity on BiPart from the two (partly overlapping in time) PB processes in Turin and Monza, managed through the same software configuration. Turin2014 was designed according to the type A approach that gives a small number of citizens the right to make and set the proposals to vote on at physical meetings. Monza adopted the type B approach, in which people have to look for followers' support to bring their proposals to a vote. The figures provide evidence that the latter energizes citizen participation. Specifically, it encourages them to bring their civic intelligence (Schuler 2001) into the process.

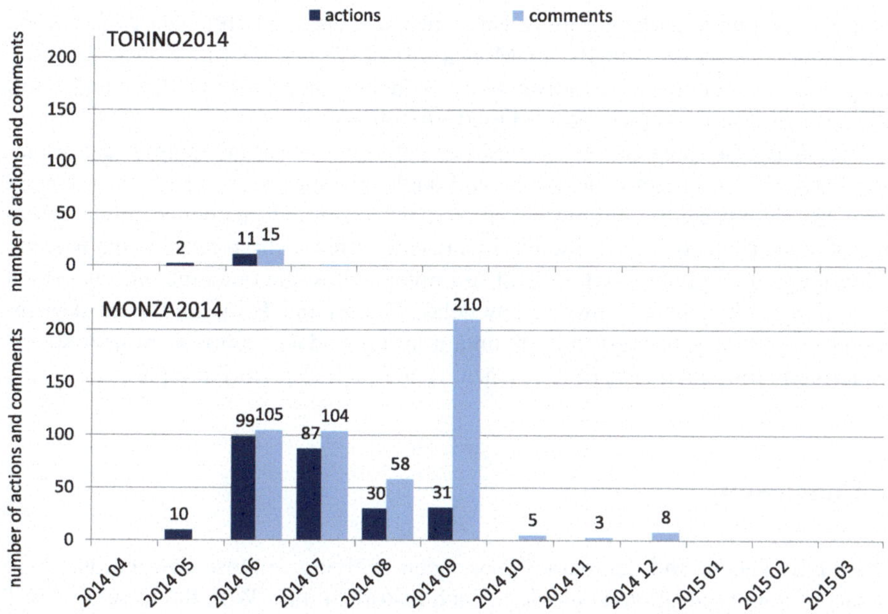

Fig. 5 Online activities in Turin2014 and Monza2014 (supports and votes are excluded)

To complete the picture of the impact of these PB processes, we must also consider what the winning projects were and what their current status is, i.e., whether municipal administrations fulfilled their commitment to finance and implement them. *Canegrate Partecipa*! gained recognition and was repeated not only because of good turnout in the two first editions but also due to the reasonable and selfless proposals that citizens first made and then selected through the deliberative process. In 2010, among seven options, people decided to refurbish the municipal sports arena. In 2011, they chose to restore the vandalized train station and build a brand-new recreation area. Both proposals came from private citizens' independent initiative, rather than from organized groups. Even more interesting, the PB process changed remarkably in 2013 (see Sect. 4), and this change reduced participation. However, this pilot process saw citizens again vote for a fully selfless project, namely "to provide support for families with financial hardship" as the city's priority, followed by a bike station near the train station.

Similar results occurred in Cernusco Lombardone, as well as in larger cities that shared parks and green spaces as some of their priorities. In Cernusco2012, citizens awarded funds to a volunteer organization for its plan to purchase new public street-cleaning equipment and voted for an urban renewal project in an outlying neighborhood. In Faenza2014, five projects were financed out of the 21 most widely supported of 96 competing alternatives. They involved two parks, a sports center, a multipurpose social space, and a public kitchen for the underprivileged. In Turin, among the three alternative urban renewal projects, each claiming the total budget, public parks got more votes than schools and public spaces for sports and cultural activities. Finally, in Monza, 40 of 62 neighborhood projects won. They range from cultural initiatives to cycle lanes, community gardens, and small urban renewal projects (see monza.bipart.it/risultati).

Except for the most recent case, Monza, all the other municipalities are somehow carrying out citizens' decisions, although in some cases the recent financial crisis has unexpectedly shrunk the municipal budget and postponed implementation of some projects, as in Cascina. In any case, what really matters is the municipality's vigorous public commitment to implementing the outcome and its efforts to explain to citizens the cause of any delay. Faenza and Turin are the most virtuous in this respect, because they maintain an up-to-date report on project status, respectively through a blog (attivafaenza.it) and on BiPart (torino.bipart.it).

6 Conclusion

The public sphere and democracy have been evolving in close correlation to the means of communication available at each historical turn. With the advent of ICT and the rise of the network society, we must expect (and envision) new forms of democracy for the future, where citizens are involved in ever more powerful ways. Cities have been laboratories for democracy through the centuries, starting from Athens, where communication took place through meetings and discussions in the agora. They are still laboratories for innovation, but the "smart city" vision—as a centralized delivery platform to enhance the quality and performance of urban services (Foth and Brynskov 2016)—is not suited to such a role. People can already produce and use data to improve their life, but they are still information providers, rather than effective decision makers. What we need are "smart communities" (Gurstein 2014), where people learn from each other and are part of policymaking.

PB gained attention as a concrete innovative democratic practice that goes in that direction because of its capacity to include people in policymaking at the urban level, taking advantage of ICT in synergy with offline venues. This chapter has outlined how PB works, reporting on nine case studies of hybrid (offline and online) PB developed in Italy over the last five years. Despite some difficulties in comparing and evaluating them, since they are inherently situated sociotechnical systems (Suchman 1987), these cases provide useful insights to term PB as an interesting and "smart" practice.

Both PB and the Internet—separately or combined—enable people to interact and aggregate fluidly (that is, directly and by entrusting others) around specific issues they care about. Both help collect civic intelligence on city needs and solutions. For instance, thanks to the georeferencing built into the software tools, proposals, and projects can visualized on a map, available to everyone's contribution. However, although the Internet broadens access to information and fosters citizens' right to have their voices heard, it does so in a crowdsourced, haphazard fashion. PB makes community members part of a deliberative process that finalizes people's contributions to a decision. It adds deliberation to participation.

The analysis of these case studies shows that the carefully designed interplay between offline and online venues is needed to make participation as inclusive as any democratic process requires. On the one hand, the online dimension overcomes the barriers to the face-to-face meetings typical of traditional PB—an underestimated physical divide—bringing participation into people's homes. On the other hand, offline equipment (e.g., mobile polling stations) offers an open window to citizens still affected by the digital divide. Offline methods dominated the earliest, smallest initiatives, probably offsetting the still embryonic online tools. However, the larger the community, the less feasible such options become, growing too expensive and ever harder to manage. In such cases, digital solutions help maintain good participation rates, even when the resources available for sustaining the PB process are scarce. This is, indeed, what happened in Cascina2013, in Faenza, in Turin, and in Monza, where online votes significantly complemented, even outpaced, those filed offline.

The Internet sustains PB, especially its scaling-up in the case of large cities. Conversely, PB encourages online participation because it motivates people to use technology. In this respect, our comparative analysis, namely the comparison between the latest cases, Turin2014 and Monza2014 (cf. Fig. 4), tells of a difference due to the choice of one participatory approach as opposed to another. Monza's PB was grounded on the *systemic approach* for which even informal discussion is part of the deliberative system because it helps shape the formal decisions. Figure 5 shows that Monza2014 succeeded in engaging citizens online to submit proposals and to debate them. In fact, not only did they populate the map with 196 proposals (61 came from the offline meetings and were uploaded by the public employees), but they also engaged with one another, providing material through 682 comments. The finalist projects all came from there. In Turin2014, the Turin District 7 PB, close to the ideal proceduralism, on the other hand, finalist projects were designed by a sample of 15 citizens. The other citizens were asked to provide input (needs and ideas) and eventually to vote on the three resulting projects. Figure 5 shows that very few people felt enough commitment to contribute online by uploading ideas and comments on the software platform.

In this context, software does matter. BiPart is an attempt to develop a full-fledged platform, to empower this democratic practice, and to make communities autonomous and smarter. It was incrementally designed and implemented to fulfill the needs of the various PB initiatives. Absolute figures for online supporters (Table 3, column 4) and voters (column 8) on BiPart, as well as statistics on the action of creating and commenting on content (Fig. 4), show a positive trend in the use of the platform, more tangible in the two latest versions. However, a dedicated usability analysis of BiPart 3.2 (Anzani 2015) points out many remaining deficiencies, which need to be fixed. For this reason, BiPart 4.0 is now being recoded within the open-source openDCN software platform (openDCN.org, where DCN stands for open Deliberative Community Networks), outlined in De Cindio and Schuler (2012).

Experience shows that PB needs to be deliberative and iterated through the years, so that citizens increasingly trust that their participation will actually be rewarded in terms of real impact on urban policymaking. If this proves the case, PB will make citizens smarter and the city a better place to live. As Coleman and Blumler (2009, p. 166) say: "Citizens of a democracy do not want to be asked their opinions [...]; they want to see an authentic relationship between [...] input and output, touching the lever and seeing the wheel turn."

References

Allegretti, G. (2012). From skepticism to mutual support: Towards a structural change in the relations between participatory budgeting and the information and communication technologies? In: P. Mindus, A. Greppi & M. Cuono (Eds.), *Legitimacy_2.0. E-Democracy and Public Opinion in the Digital Age* (pp. 145–182). Frankfurt am Main: Goethe University Press.

Allegretti, G., & Dias, N. (2015, in press). Instituições sem edifícios: os orçamentos participativos numa perspetiva portuguesa. Coimbra: Almedina (in Portouguese).

Allegretti, G., Sintomer, Y., Herzberg, C., Röcke, A., & Lopes Alves, M. (2013). Participatory budgeting worldwide—updated version, dialog global, 25. http://www.service-eine-welt.de/en/images/text_material-3651.img. Accessed 15 June 2015.

Allegretti, G., & Stortone, S. (2014). Participatory budgets in Italy: Reconfiguring a collapsed panorama. In: N. Dias (Ed.), *Hope for democracy—25 years of participatory budgeting worldwide* (pp. 313–324). In Loco.

Antoniadis, P., & Apostol, I. (2014). The right(s) to the hybrid city and the role of DIY networking. *The Journal of Community Informatics*, *10*(3). Retrieved from http://ci-journal.net/index.php/ciej/article/view/1092.

Anzani, S. (2015). User eXperience in casi di bilancio partecipativo online. BA Dissertation in Digital Communication, Università degli Studi di Milano, AA 2013/2014, (in Italian).

Aurigi, A., & De Cindio, F. (2008). *Augmented urban spaces: Articulating the physical and electronic city*. Aldershot: Ashgate.

Avritzer, L. (2009). *Participatory institutions in democratic Brazil*. Pennsylvania: Woodrow Wilson Center Press.

Bertone, G., De Cindio, F., & Stortone S. (2015). LiquidFeedback in large-scale civic contexts: Framing multiple styles of online participation. In: G. Convertino, A. Westerski, P. Diaz & A. De Liddo (Eds.), Special issue on large-scale ideation and deliberation: Tools and studies in organization. *The Journal of Social Media in Organizations*, *2*(1). Retrieved from http://www2.mitre.org/public/jsmo/pdfs/02-01-liquid-feedback.pdf.

Bobbio, L. (2010). Types of deliberation. *Journal of Public Deliberation*, *6*(2), Article 1. Retrieved from http://www.publicdeliberation.net/jpd/vol6/iss2/art1.

Bordignon, F., & Ceccarini, L. (2013). Five stars and a cricket. Beppe grillo shakes Italian politics. *South European Society and Politics*, *18*(4), 427–449.

Castells, M. (2009). *Communication power*. Oxford: Oxford University Press.

Castells, M. (2012). *Networks of outrage and hope: Social movements in the internet age*. Cambridge: Polity Press.

Coleman, S., & Blumler, J. G. (2009). *The internet and democratic citizenship: Theory, practice and policy (communication, society and politics)*. Cambridge: Cambridge University Press.

De Cindio, F. (2012). Guidelines for designing deliberative digital habitats: Learning from e-Participation for open data initiatives. *The Journal of Community Informatics*, *8*(2). Retrieved from http://ci-journal.net/index.php/ciej/article/view/918/910.

De Cindio, F., & Schuler, D. (2012). Beyond community networks: From local to global, from participation to deliberation. *The Journal of Community Informatics, 8*(3). Retrieved from http://ci-journal.net/index.php/ciej/article/view/908/933.

Dias, N. (2014). *Hope for democracy—25 years of participatory budgeting worldwide* (pp. 413–425). In Loco.

Foth, M. & Brynskov, M. (2016, in press). Participatory action research for civic engagement. In: E. Gordon & P. Mihailidi (Eds.), *Civic media: Technology, design, practice.* Cambridge: MIT Press.

Foth, M., Forlano, L., & Satchell, C. & Gibbs, M. (Eds.). (2011). *From social butterfly to engaged citizen: Urban informatics, social media, ubiquitous computing, and mobile technology to support citizen engagement.* Cambridge, MA: MIT Press. ISBN 978-0-262-01651-3.

Gurstein, M. (2014). Smart cities vs. smart communities: Empowering citizens not market economics. *The Journal of Community Informatics, 10*(3). Retrieved from http://ci-journal.net/index.php/ciej/article/view/1172.

Iglesias Turrion, P. (2014). *Disputar la Democracia. Politica para tiempos de crisis.* Madrid: Akal.

Mansbridge, J., Bohman, J., Chamber, S., Christiano, T., Fung, A., et al. (2012). A systemic approach to deliberative democracy. In J. Parkinson (Ed.), *Deliberative systems.* Cambridge and New York: Cambridge University Press.

Matheus, R., Ribeiro, M., Vaz, J. C., & de Souza, C. A. (2010). Case studies of digital participatory budgeting in latin America—Models for citizen engagement. In J. Davies & T. Janowski (Eds.), *Proceedings of the 4th International Conference on Theory and Practice of Electronic Governance* (pp. 31–36). New York: ACM Press.

Peixoto, T. (2009). E-Participatory budgeting: e-Democracy from theory to success? *European Journal of ePractice., 7,* 1–9.

Sampaio, R. C. (2014). e-Orçamentos Participativos: um mapeamento mundial das experiências já realizadas e suas contribuições para e-participação e e-democracia, Universidade Federal da Bahia. PhD dissertation in Social Communication (in Portuguese).

Sampaio, R. C., & Peixoto, T. (2014). Electronic participatory budgeting. False dilemmas and true complexities. In: N. Dias (Ed.), *Hope for democracy—25 years of participatory budgeting worldwide* (pp. 413–425). In Loco.

Scherer, S. & Wimmer, M. A. (2012). Reference process model for participatory budgeting in Germany. In: E. Tambouris, A. Macintosh, & Ø. Sæbø (Eds.), Electronic Participation, Proceedings ePart 2012, Lecture Notes in Computer Science. (Vol. 7444, pp. 97–111). Berlin/Heidelberg: Springer.

Schuler, D. (2001). Cultivating society's civic intelligence: Patterns for a new "world brain". *Information, Communication and Society, 4*(2), 157–181.

Shah, A. (Ed.). (2007). *Participatory budgeting.* Washington: The World Bank Press.

Sintomer, Y., Herzberg, C., Röcke, A., & Allegretti, G. (2012). Transnational models of citizen participation: The case of participatory budgeting. *Journal of Public Deliberation, 8*(2). Article 9. Retrieved from http://www.publicdeliberation.net/jpd/vol8/iss2/art9.

Sintomer, Y., Traub-Merz, R., & Zhang, J. (Eds.). (2013). *Participatory budgeting in Asia and Europe. Key challenges of participation.* Basingstoke and New York: Palgrave Macmillan.

Stortone, S. (2010). Participatory budgeting: Towards a "civil" democracy? In: M. Freise, M. Pyykkönen, & E. Vaidelyte (Eds.), *A Panacea for all Seasons? Civil society and governance in Europe.* Baden-Baden, Nomos.

Suchman, L. (1987). *Plans and situated actions. The problem of human-machine communication.* New York: Cambridge University Press.

Open Government: Transforming Data into Value-Added City Services

Agnes Mainka, Sarah Hartmann, Christine Meschede and Wolfgang G. Stock

Abstract Open data and open government are topics which are not new, but today we can see initiatives which transform data into value-added services. But are there real world examples where those services make the city "smart?" We are going to define open urban government data in the context of e-Government and m-Government. This chapter reports on an investigation about the open urban government data and hackathon movement on a global scale and takes a deeper look at the real world examples of Amsterdam, Barcelona, and Paris. It will be discussed which challenges governments are confronted with to make open data available as well as the role of hackathons in the development of mobile applications based on this data.

Keywords Open data · Open government · e-government · m-government · Hackathons · Smart city

A. Mainka (✉) · S. Hartmann · C. Meschede · W.G. Stock
Department of Information Science, Heinrich-Heine-University Düsseldorf, Düsseldorf, Germany
e-mail: agnes.mainka@hhu.de

S. Hartmann
e-mail: s.hartmann@hhu.de

C. Meschede
e-mail: christine.meschede@hhu.de

W.G. Stock
e-mail: stock@phil.hhu.de

M. Foth et al. (eds.), *Citizen's Right to the Digital City*,
DOI 10.1007/978-981-287-919-6_11

1 Introduction

Establishing open data in the government sector was recognized in the European Union (EU) since the end of the 1980s (Janssen 2011). So the concepts of open government and open data are not new, but with the development of ICT and digitization we arrive at new possibilities which may result in value-added data or services. These concepts aim at enhancing transparency, participation, and collaboration between the public and political agencies (McDermott 2010; Lathrop and Ruma 2010). In 2003, the EU introduced the Public Sector Information Directive (van der Waal et al. 2014). In this case, restrictions for using and reusing public data have been eliminated, but the data access was charged by governments. The pricing was calculated as cost covering, but this is not really "open data." Today, we understand open data as (online) accessible data which is free of charge or with just marginal costs of use (Jetzek et al. 2013). When we talk about open data, which is generated by the government, we can refer to "open government data" [OKF (date unknown)]. Until now, we can find a lot of open government data initiatives on the national level, e.g., data.gov (USA), data.gov.uk (UK), on the international level open-data.europa.eu (European Union Open Data Portal), and on the city level open.wien.at (Vienna, Austria), or nycopendata.socrata.com (New York City, USA). To open government data causes a lot of challenges which need to be faced. In addition, it is not clear whether these efforts result in a more open, transparent, and collaborative government.

Just opening datasets is not the whole concern of open government; they also like to animate citizens, startups, and other stakeholders to reuse the data. For this reason, workshops or hackathons (hack marathons) are organized with the goal to reuse open government data collectively (Baraniuk 2013; Briscoe and Mulligan 2014). Here, we can differ between top and down initiatives headed by government agencies, e.g., "apps4austria" in Austria, or bottom-up run by engaged citizens, e.g., the "Open City" volunteers in Chicago. In many cases, the output of those hackathons is mobile applications (m-apps) which intent to solve urban problems. Hackathons have established themselves as common solutions for building m-apps based on open government data, but the outcomes attract just little attention in academic research (Johnson and Robinson 2014). Thus, we do not know whether the output of hackathons results in a value-added city service.

The chapter is structured into five parts. After the introduction, the aim of open government data initiatives and our research approach is presented (Sect. 2). This will be followed by the used methods (Sect. 3) and the results of interviews conducted with experts from politics and universities in Paris, Amsterdam, and Barcelona (Sect. 4.1) as well as an investigation of the hackathon movement and output on a global scale (Sect. 4.2). In Sect. 4.3, the open data initiatives and hackathons of the three cities (Paris, Amsterdam, and Barcelona) are analyzed. Finally, the chapter ends with the conclusion (Sect. 5).

2 Open Government Data Initiatives

In our context, *data* means quantitative values resulting from measurements and other sources. *Open data* refers to data, which is freely accessible online, while there are no technical or legal restrictions to reuse such data (Jetzek et al. 2013). *Open government data* is defined as "open data produced by the government" [OKF (date unknown)] as a subset of public sector information. Finally, *open urban government data* is open government data on the municipal level.

There are two types of sources for open urban government data: official statistics and sensor-based data. Official statistics include data on population, business and economics, jobs, crimes and justice, and health. Additionally, there are city-specific official data collections such as the urban forest map of San Francisco with detailed data about trees in the city. Sensor networks (Kitchin 2014) consist of sensors which are embedded in specific structures and measure—in real time—levels of light, humidity, temperature, air pressure, movements, speeds, etc., and of transponders monitoring empty spaces in car parks, data from closed circuit television (CCTV), or the progress of trains and buses along a route (Kuhn 2011). All in all, sensor-based "big data" with relevance for the city (Bettencourt 2014).

This open urban government data can be combined with user-generated content, for example, GPS-based data from mobile devices and posts on micro-blogging services, such as Twitter, since cities' e-governments allow citizens to give feedback on mobile or Web applications and use social media channels (e.g., Facebook, YouTube, Flickr; Mainka et al. 2015a; Mergel 2013). It is also possible that private companies open parts of its data, e.g., on the companies' products or services with relation to the city (Immonen et al. 2014).

Combining these data, we see many services programmed by governmental agencies, by private developers (Mainka et al. 2015b) or which evolved from hackathons (Johnson and Robinson 2014). Reviewing the literature of the last decades, not everybody could imagine that citizens would act as developers for governmental services—*without being paid for it.* What we see today is a game between open data support by governments and data reuse by residents and other stakeholders. Its gain is called public, social, or economic value and an enhanced quality of life. Public value is in our history often a reason for volunteering, for example, citizens acting as bodyguards in neighborhood safety patrols. Citizens may "provide more and better services than they would have received" by their government (Bellone and Goerl 1992, p. 134). Today, we have a kind of *data bodyguards*, who help themselves and their community to handle open data and transform it into value-added services, for example, smartphone apps. Nevertheless, the new possibilities arriving with ICT and digital open data confront us with different challenges which need to be faced. In the Apps for Smart Cities Manifesto, we read, "to harness the true potential of Smart cities, the city must become a platform i.e. an enabler for developers, creativity and applications. In doing so, the city becomes like the internet i.e. a connector and an enabler for citizens which aims to empower the citizen" (Apps for Smart Cities 2012). Urban m-apps are indicators for the evolution of “ubiquitous government” (Belanger et al. 2005) or “smart government.” “Smart Government will share resources and information and interoperate with other governments, citizens, NGOs and for-profit businesses much more smoothly than today” (Scholl 2012, p. 324).

The shared resources include “big data” as well as “accurate, comprehensive, and reliable information” (Scholl and Scholl 2014, p. 167). Open urban government data shows its assets best in combining it with further open data and open tools such as OpenStreetMap.org in the sense of mashups. In addition, urban m-apps include location-based participation, “for example helping to redesign the park you're walking in, or the hospital organization which kept you waiting and you think you have a solution” (Millard 2010, p. 8). But citizens' participation goes far behind the simple commenting of communal problems and includes the production of urban m-apps as well. Desouza and Bhagwatwar (2012) call such applications to solve even complex urban problems *citizen apps*.

Reviewing the ongoing research on governmental development in association with information and communication technology, diverse terminologies are used. In our context, m-government is the use of mobile technologies (e.g., mobile phones, smartphones, and tablets) in e-government. M-government enables location-based services (Carroll and Ganoe 2009), which are “personalized services delivered to a mobile device user at a remote location.” To make open data reusable, standards have already been established, e.g., by a W3C eGovernment Interest Group (www.w3.org/egov). They should help the government to open their data and other stakeholders to reuse this data. Concerning OECD and ITU (2011), we are in need of application developers (programming the interfaces between the device and the network) as well as content developers and enablers (compiling content into mobile-ready formats). Hence, citizens become “data prosumers (both consumers and providers of data)” (Charalabidis et al. 2014). Thus, “factors as the cost and availability of Internet access, the language in which the data is presented, the technical or professional requirements for interpreting and making use of the data, and the availability of training in data use and visualization, among others” are important for the success of open government data (Gurstein 2011). Additional factors are the adoption by citizens in their mobile phone use, social factors, income level, and trust of service, but not demographic factors (Liu et al. 2014; Reddick 2014; Venkatesh et al. 2014). The critical success of an m-government service or m-app is the service's perceived usefulness (Hung et al. 2013).

The development of urban m-apps should be embedded in an “open data based business ecosystem” (Immonen et al. 2014, p. 88), which consists of elements such as key partners, co-creation, revenue strategies, customers and markets, data structure, business development, and the value added (Immonen et al. 2014, p. 92). The objective of a business ecosystem is to ensure the sustainability of m-apps (and not a quick development of an app followed by stopping the product some weeks later). Figure 1 gives an overview on issues of developing urban m-apps.

To build m-apps based on open urban government data in order to enhance collaboration, quality of life, and to solve urban problems is a common topic in many cities around the world (Mainka et al. 2015a). Some cities are fast adopters of this

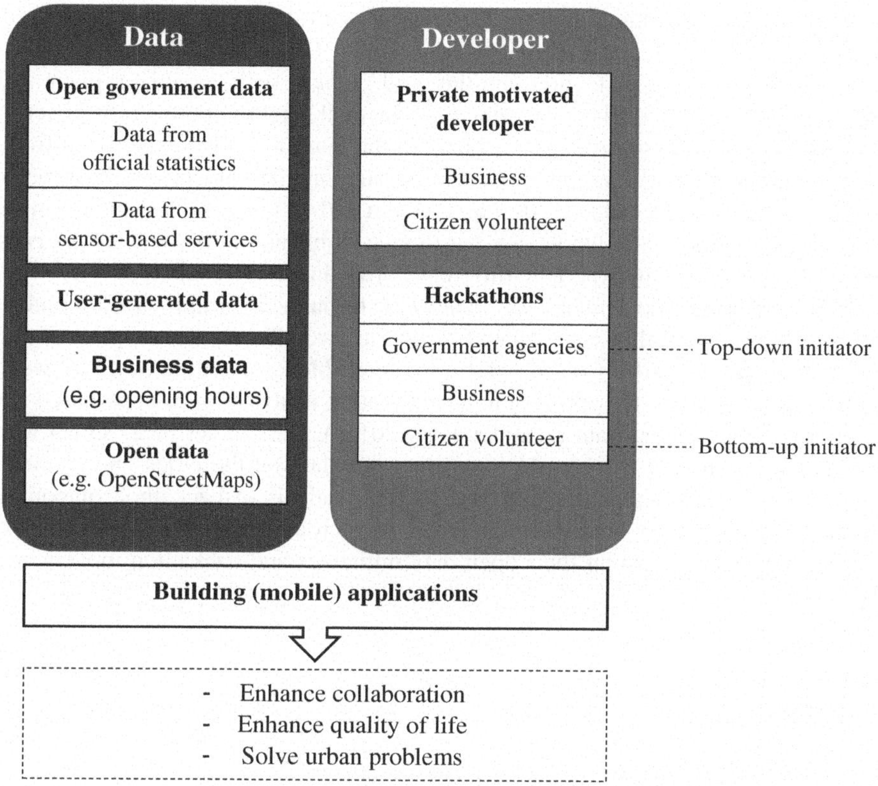

Fig. 1 Entities related to the development of m-apps based on open urban government data

approach, and other cities are latecomers or still fear the challenges which emerge with open government data initiatives (Janssen 2011). Other cities start hackathons without certainty of its usefulness (Johnson and Robinson 2014). Since there is just little research on this development, we want to locate the challenges of opening data and analyze in a case study the cities Paris, Barcelona, and Amsterdam. Therefore, we interviewed persons from politics and research to get insights about the current development and its obstacles. Furthermore, we analyzed the hackathon movement in 24 cities to identify best practice examples and detect whether these hackathons help governments to overcome fears and barriers.

3 Methods

For the investigation of the challenges of open government, we choose expert interviews, because this method is appropriate "for an orientation in the field" (Flick 2009, p. 528). In our approach, the interviews are used to identify the challenges this development poses. The interviews are based on a semi-structured method with open and closed questions (Creswell 2012).

The interviewed experts are familiar with topics such as urban planning, e-government, public institutions, and ICT. In total, eight experts from government agencies and universities participated in the interview. Two experts are from Barcelona, two from Paris, and four from Amsterdam. The interviews have been conducted between December 2013 and March 2014. The results do not aim at reflecting the opinion of a whole city but to give an insight into how *open government* is understood from experts in different European cities (result Sect. 4.1).

The hackathon movement was analyzed through a simple Web content research. Since hackathons are related to online available resources, promotions and reviews are hosted on online platforms, e.g., blogs or online news as well. Additionally, we refer to a prior conducted content analysis on m-apps based on open urban government data (Mainka et al. 2015a). Here, a set of 24 cities was analyzed by searching for m-apps based on open data on the cities' governmental Web sites. These cities are our base for the analysis of hackathon outcomes around the globe (result Sect. 4.2). The cities Amsterdam, Barcelona, and Paris are deeper analyzed concerning their open data initiatives and hackathon movements (Sect. 4.3).

4 Results

4.1 Challenges of Open Urban Government Data

Today, we see a lot of barriers and opportunities to make open urban government data accessible. In this section, we want to take a deeper look at current barriers which hamper governments to open their data. In their study, Huijboom and Van Den Broek (2011) have mentioned, for example, a "closed government culture …, privacy legislations …, confidentiality, risk avoidance and fear of political escalation" as the main reasons. To better understand where these fears and unwillingness come from, we asked the experts which challenges they define as the biggest within the open government trend. According to a prior literature review, the challenges were structured into the following six categories:

1. *Political Challenge*: Politicians fear to lose their monopoly in public affairs. To shift from a close to a more open government, "a deep reform of public administration ways of thinking and behavioral patterns, organizational structures, and operational approaches" are needed (Moro 2005, p. 110).
2. *Legal Challenge*: Security reasons, privacy, and copyright are often used as arguments to protect data from the public (Huijboom and Van Den Broek 2011).

3. *Governance Challenge*: Governments' motivation to open their data and to collaborate includes the participation of business and citizens as well (Huijboom and Van Den Broek 2011). In addition, it is questionable in which form the data should be opened.
4. *Human Resource Challenge*: Diverse skills are needed from government and stakeholders. It is a highly complex task to prepare the data to be reused in a variety of ways (Dawes and Helbig 2010). Government agencies are in need of talents who are able to prepare the data to be open and reusable.
5. *IT Infrastructure Challenge*: Information and communication technology is the essential driver of digital open data. Thus, it is considerable for a government to be connected through an enhanced IT infrastructure to make online services available to the public.
6. *IT Budget Challenge*: How governments spend their funds may differ in every country and depends on their overall available budget. The financing plays a role especially when governments have implemented charging models for their data (Huijboom and Van Den Broek 2011). To open those data will entail a financial lack.

The literature review points out that open government is not a new idea, but becoming an open government is still in process. The experts were asked to indicate which of the mentioned challenges they perceive as the main challenges to the provision of public data as open data by their (1) municipality and (2) central government. Additionally, a brief motivation of their decision was given by the interviewees. The experts had to choose two challenges which they see as the main ones. For the most interviewees, the political challenge was identified as one of the biggest challenges. It could also be described as a "strategic challenge." If they open data, they can lose the control about urban services. In Web 2.0 services, the users provide the information, which is a very important development. Accordingly, "the government won´t be the monopolist of data" (Alain Rallet, 11 December 2013, Paris). Another aspect is the priority of this topic within the municipality. "Open data is not one of the most important topics for the politics" (Marta Carrasco Bonet, 4 December 2013, Barcelona). However, one of the major problems in the political challenge could be the fear to lose power. "The politicians think that information is power and they do not want to lose this power. They are not aware of the fact that cross information makes sense. Not just open data but also internal data sharing will improve a lot" (Joan Batlle-Monterrat, 4 December 2013, Barcelona).

As a legal challenge, most experts mentioned privacy rights. It is not clear who will be the owner of open data and who will be accountable for them. Furthermore, the experts are consistent in their concerns about the governance challenge. The data needs to be collected and shared in a reusable form which is a very big issue. The challenges "human resource," "IT infrastructure", and "IT budget" are deeply related to each other. Thus, depending on the budget, talents can be hired who implement and maintain the IT infrastructure. Unlike in the literature, where the IT infrastructure is a challenge for suburbs or not connected regions, the experts live and work in cities with enhanced IT infrastructures. Therefore, they associate this question with the internal IT infrastructure of their government. The IT infrastructure is especially a challenge in Amsterdam where "the national government had several 'disaster' in IT infrastructure projects in the past few years" (Marco Bontje, 21 January 2014, Amsterdam).

If we take a look at the main challenges for the municipal government and the central government, there are just marginal differences. It highly depends on the political structure in each city and nation. If the central government is accountable for the IT budget, then IT budget could not be a challenge for the municipality and vice versa. In the case of Amsterdam, we can see that the experts count the IT infrastructure not just as a challenge for the central but also for the municipal government.

4.2 The Hackathon Movement

In the last few years, a new trend has appeared on how governments and citizens work together to reuse open urban government data and build useful services with it. In order to realize this, events are hosted which are named "hackathon," "app contest" (Johnson and Robinson 2014), or "app competition." But what is it all about? Do hackathons or similar contests help to develop value-added services? To find evidence about these aspects, we analyzed hackathons in 24 cities, which are prototypical cities of the knowledge society (Mainka et al. 2015a). First, we will specify the types of those events and their aim, and identify the participants. According to a prior research on m-apps based on open urban governments (Mainka et al. 2015a), we will investigate if and how these hackathons are related to successful m-apps based on open urban government data.

The term hackathon combines the two words "hack" and "marathon" which is also a hint at the temporal construction of those events, in which the term "hack" refers to smart programs or programmers and not to criminal hack attacks. In general, those events are often hosted to solve specific problems or to address stated topics, e.g., tourism and culture, democracy, mobility, security, energy, waste, transportation, and water efficiency (Apps for Amsterdam 2012; Cleanweb Worldwide 2012). The idea is to bring people from different backgrounds together and see whether they are able to create value-added products. The initiative can come from the government (top-down) as well as from the hacker community (bottom-up). However, the aim to host such events is mostly to build a bridge between government, citizens, and economy. Vienna stated that the citizens are the most needed and emphasize that especially statisticians as well as librarians are wanted participants (Barcamp 2013). There is no limitation on the participation level, e.g., the "apps4austria" competition invites companies, institutions, and administrative organizations to participate as well (Digitales Österreich n.d.). But not only governments and citizens organize hackathons, there are also new companies that emerge to host these events. One example is angelhack which is operating since 2011 and organizes hackathons for other companies, e.g., Adobe in San Francisco (AngelHack 2011).

Hackathons are often hosted in and by cities and regions all over the world, e.g., Amsterdam, as well as for bigger areas, e.g., the whole country, such as in Finland (Apps4Finland) or the whole European Union (Apps for Europe n.d.). However, it has its origins in the USA where the first hackathon has been held in 2011 (Headd 2011).

Another aspect which can differ between events is whether the hackathon is awarded or just for fun. Furthermore, which status do the developed applications or services have. Are prototypes ready for use? In some cases, ideas to solve problems are highly welcomed and awarded, e.g., for the Stockholm Award. This award is part of the "Vision 2030, where Stockholm will be a vibrant, humane and creative world-class city, the competition was to bring the City a step in the right direction" (Studentcompetitions 2012).

If the resulting ideas or applications will bring a social or economic value after the hack, will be decided by the market. The ideas and products have to be accepted and used. In some cases, the developer gets some assistance by the city or the initiators of the hackathon. This can be financial or technical aid, or arrangements in order to connect the developer with existing project groups. An example of "crowdfunding" is the platform sf.citi by San Francisco. They try to support developers and products by organizing fund through the public. For example, citizens can fund the projects they prefer on platforms like "citizenvestor" (Hunter 2013).

The challenge is not to develop a m-app, but to make it successful. Therefore, the products of civic hacking have to be promoted after these events so that they get recognition by a larger quantity of citizens. For this purpose, Chicago provides a good example. They call for residents to become a member of "The Civic User Testing Group" (CUTGroup) which is a project by the "Smart Chicago Collaborative."

The increasing popularity of open data also changes the quality of the data. Now, governments see that the data is used and they try to face the new challenges. Thus, new jobs need to be created as well; for example, a Chief Data Officer was proved for the city of San Francisco (Hunter 2013). Hackathons and app challenges have even become so popular that guides, e.g., "How to run a hackathon"[1] and books, e.g., "Civic Apps Competition Handbook"[2] were published to support hosting these events.

---

[1]http://www.socrata.com/wp-content/uploads/2013/05/How-to-Run-a-Hackathon.pdf.

[2]http://govfresh.com/2012/12/civic-apps-competition-theres-a-book-for-that/.

4.3 Open Government Data Initiatives and Hackathons in Amsterdam, Barcelona, and Paris

The biggest challenge to become an open government mentioned by the interviewed experts is the political challenge (see Sect. 4.1). The cities Amsterdam, Barcelona, and Paris have already started to open their urban government data online. Of course, there are differences in the amount of open data sets and addressed topics, but strategies and plans to join open data initiatives are already transformed into real action. Amsterdam supports a metadata platform (www.amsterdamopendata.nl), which links to the open datasets of available geo data (maps.amsterdam.nl/open_geodata) and statistical data (www.ois.amsterdam.nl/feiten-en-cijfers) of different government institutions. In contrast, the open data portals from Barcelona and Paris host the datasets on one portal. All cities use different licensing types, e.g., creative commons or own licenses. They have in common that the data is allowed to be reused in further projects, e.g., in new applications. All cities mention in their terms of use that the city is the owner of the data but is not responsible for the accuracy of it. The major part of the data is available in machine-readable formats, e.g., CSV, JSON, or ODATA, and for geographical data in Shapefiles (SHP) or GeoJSON. Amsterdam and Barcelona also offer PDF files or even images. Some of these "datasets" are scanned documents, e.g., administrative bylaws, which are not machine-readable. Table 1 gives an overview of the open data portals, datasets, data formats, and the categorized themes of data for each city.

The interviewed experts see that human resource, IT budget, and IT infrastructure are related topics. Barcelona, for example, has established new jobs which are responsible for the open government movement in their city. These positions are called for instance "Smart City Director & Deputy CIO" or "Mobile, e-Government & Data Director." In contrast, Paris follows a defined action plan which is dedicated to the modernization of public action directed by the prime minister of France. A particular institution was founded, called Etalab (www.etalab.gouv.fr), which is responsible for the national open data development. In Amsterdam, the Amsterdam Smart City platform was founded to bring partners from business, research, government, and citizens together.

Barcelona's open data initiatives are based on tools developed by Microsoft (Microsoft 2014). Thus, they have datasets, e.g., ODATA, which are mostly used in Microsoft environments. In addition, the governmental internal IT infrastructure is based on Microsoft products. Whether Barcelona receives benefits from Microsoft for using their tools and operating systems is not clear. They also collaborate with Cisco concerning their smart city initiatives, e.g., implementing sensors in the city (Cisco 2011). In addition, the cooperation in research and technology between IBM and the Barcelona Supercomputing Center helps the city to become a vanguard of open government (IBM 2015). To adapt m-government services, citizens and tourists may use one of the 700 Wi-Fi hot spots all over the city. Amsterdam also collaborates with companies to improve its open data initiatives. Together with the navigation system company TomTom, they built an open

Table 1 Open datasets in Amsterdam, Barcelona, and Paris (retrieved June 2015)

• **City** • Amount of open datasets • Open data portal • License	Data format	Themes
• **Amsterdam** • 424 • www.amsterdamopendata.nl • Own license (data is allowed to be used and re-used for commercial and non-commercial purpose. City of Amsterdam has to be cited as source holder)	(Images) PDF XLS CSV GeoJSON GeoRSS SHP	Management and organization Population Services Economics and haven Education, youth and diversity Energy Geography Environment and water Public order and safety Public space and green Sport and recreation Urban development Tourism and culture Traffic and infrastructure Elections Work and income Housing and environment Care and welfare
• **Barcelona** • 323 • opendata.bcn.cat • Creative commons licenses 3.0	PDF CSV XML ODATA	Administration Economy and business Population Territory Urban environment
• **Paris** • 130 • opendata.paris.fr • Own license inspired by the open database license (ODbL) v1.0 (data is allowed to be used and re-used under share-alike conditions)	CSV XLS JSON GeoJSON SHP	Culture Urban planning Citizens Environment Movement Service

dataset to detect free parking slots in the city (Balch 2013). This community-driven approach is also reflected at their open data portal where the Amsterdam Economic Board, the Waag Society, the Vrije Universiteit, and the Universiteit van Amsterdam work together. In Paris, the open data initiatives are inspired by the Open Knowledge Foundation. This foundation is a non-profit organization and supports others, e.g., governmental organizations, to join in the open data movement. Direct corporations on the city level as in Barcelona could not be detected in Paris, but national government institutions work together with OpenDataSoft (www.opendatasoft.com) in France.

Having all the data online allows these cities to run own hackathons which may reuse open urban government data. The majority of the detected hackathons are initiated by NGOs, universities, governments, or companies. In Amsterdam, we can find hackathons such as the Open Data FWD, which is a project of the

Department of Infrastructure, Traffic and Transportation (DIVV) of the City of Amsterdam, and Waag Society, or Apps for Amsterdam which focuses on getting Open Data apps on the market and is organized by the Waag Society in corporation with the Economic Development Board Amsterdam (EDBA). In Barcelona, no hackathon could be found which is dedicated to work on urban government data. Nevertheless, a hack community is present in the city, and other hackathons such as Hackathon for Social Good, Journalism Hackathon, or A Smart Cities Hackathon are held there. Also, workshops for hackers are offered during the Barcelona Smart City App Hack. Bottom-up initiatives are hosted by a community called the Databeers. They arrange meet-ups at irregular intervals. Equally in Paris, no open urban government data hackathon could be located. However, national or commercial hackathons are hosted in Paris, too.

Thus, the questions arise as follows: Are open data initiatives and hackathons useful to build value-added city services? In the case of Amsterdam, we have to answer "yes." They have a lot of open data available and host hackathons. For example, the Apps for Amsterdam hackathon in 2012 has announced the "Parkshark API" and "Bike like a Local" as winners of their contest. In the case of the Parkshark API, not only the city of Amsterdam can benefit from it, because less cars drive through the streets searching a parking slot which helps to reduce carbon emissions. The citizens are winners as well, since they now can find cheaper and faster parking slots nearby. But also the developers of the m-app benefit of this competition. Their product has become famous which has led to commercial orders.

Our investigation has brought to light that most hackathons address people who are programmers or designers, which is not in common with the cities open data vision. Thus, the majority of the citizens will not feel affiliated to join hackathons or app competitions. Acknowledging our prior study on m-apps built on open urban government data (Mainka et al. 2015a), m-apps developed by NGOs, private companies, or citizens have been produced in all three cities. In some cases, the cities' government emerged as the developer of some m-apps. Figure 2 illustrates

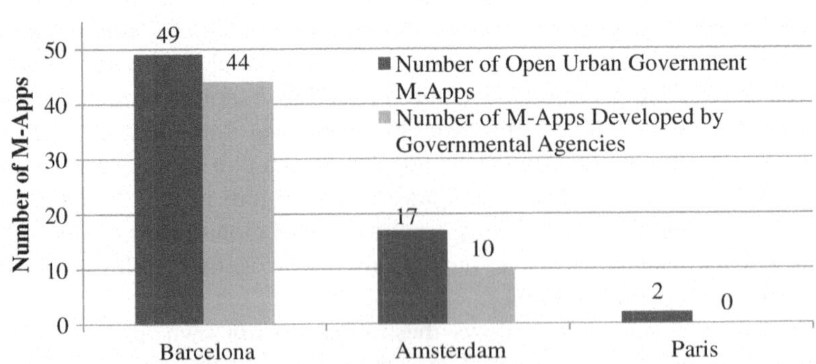

Fig. 2 M-apps found on governmental Web sites of Amsterdam, Barcelona, and Paris as well as amount of m-apps developed by governmental agencies

the total amount of m-apps which are based on open urban data and found on governmental Web sites as well as the amount of these m-apps developed by a governmental institution. However, in the case of Paris, most of the m-apps could not be identified as being based on open data, wherefore they had to be excluded.

5 Conclusion

Is the government able to transform data into value-added city services? Have we already arrived at an open government stage? As reviewed in the literature, the initiatives and ideas that governments should be more open and citizen oriented are not new. But a new development is the reuse of open urban data in mobile applications.

A lot of open data could be or already is available, such as sensor-based data in car parks or GPS information to locate traffic jams. That this data is reused is evidenced by the detected m-apps based on open urban government data and the burgeon hackathons around the world. Considering the challenges of opening data such as the governance challenge, IT budget and infrastructure challenge, or legal challenge, one could think that we are at the beginning of a new movement. However, the maturity of this development depends on the respective city. Barcelona, for example, has opened its data under creative commons licenses to the public. Now citizens, NGOs, and anybody else may use this data.

In some cases, we see that the initiatives of opening the data or initiating a hackathon may come from the government, in others from the community. As the first hackathons from San Francisco brought out some helpful applications, others started to adapt this idea. Today, we can find app challenges in almost every bigger city. This movement is developing and changing rapidly. Some m-apps do not exist anymore, others have been further developed and updated, and some others have not been changed or improved for a long time (e.g., Play Amsterdam: last update 06/06/11).

However, it still remains unclear whether hackathons are indeed able to make a difference and bring the citizens into the governmental processes. What we actually see is that people come together as a community and develop governmental services. This could help to make the whole city and its residents smarter.

Finally, comparing all analyzed cities, we see major differences in handling open urban government data. The hackathon movement and the availability of citizens' apps are common. There are many different kinds of hackathons and competitions, which are also co-operating with each other. Every day a new event could be hosted and others are not organized anymore. Of course, not all applications, projects, and ideas can achieve success, but more important than launching successful applications is to build a collective community that aims at solving problems together, regardless whether they are city officials or ordinary people. The main aim is to transform government, to reuse open data, and to open the data for anybody. In fact, this can be achieved with the help of hackathons and app competitions. Above all, producing useful services from which everybody will benefit should be forwarded in future.

References

AngelHack. (2011). *What is AngelHack?* www.angelhack.com/about-us/. Accessed November 28, 2014.

Balch, O. (2013). *Can open data power a smart city revolution?* www.theguardian.com/sustainable-business/open-data-power-smart-city. Accessed June 2, 2015.

Baraniuk, C. (2013). The civic hackers reshaping your government. *New Scientist, 218*(2923), 36–39. doi:10.1016/S0262-4079(13)61625-5.

Belanger, F., Carter, K. D., & Schaupp, L. C. (2005). U government: A framework for the evolution of e government. *Electronic Government: An International Journal, 2*(4), 426–445.

Bellone, C., & Goerl, G. F. (1992). Entrepreneurship democracy. *Public Administration Review, 52*(2), 130–134.

Bettencourt, L. M. A. (2014). The uses of big data in cities. *Big Data, 2*(1), 12–22. doi:10.1089/big.2013.0042.

Briscoe, G., & Mulligan, C. (2014). *Digital innovation: The Hackathon phenomenon*. CreativeWorks London Working Paper No. 6. London: CreativeWorks.

Carroll, J. M., & Ganoe, C. H. (2009). Supporting community with location-sensitive mobile applications. In M. Foth (Ed.), *Handbook of research on urban informatics: The practice and promise of the real-time city* (pp. 339–352). Hershey: IGI Global.

Charalabidis, Y., Loukis, E., & Alexopoulos, C. (2014). Evaluating second generation open government data infra-structures using value models. In *Proceedings of the 47th Hawaii International Conference on System Sciences, IEEE Computer Society, Washington* (pp. 2114–2126).

Cisco. (2011). *Press release*. newsroom.cisco.com/press-release-content;jsessionid=01230049AD3CA4E7D7A1C6438DD16818?type=webcontent&articleId=5918850. Accessed June 2, 2015.

Creswell, J. W. (2012). *Qualitative inquiry and research design: Choosing among five approaches*. Thousand Oaks: Sage.

Dawes, S. S., & Helbig, N. (2010). Information strategies for open government: Challenges and prospects for deriving public value from government transparency. In M. A. Wimmer, et al. (Eds.), *EGOV 2010, electronic government. Lecture notes in computer science* (pp. 50–60). Berlin: Spinger.

Desouza, K. C., & Bhagwatwar, A. (2012). Citizen apps to solve complex urban problems. *Journal of Urban Technology, 19*(3), 107–136. doi:10.1080/10630732.2012.673056.

Flick, U. (2009). *An introduction to qualitative research*. Thousand Oaks: Sage.

Gurstein, M. B. (2011). Open data: Empowering the empowered or effective data use for everyone? *First Monday, 16*.

Headd, M. (2011). *Open government hackathons matter*. govfresh.com/2011/08/open-government-hackathons-matter/. Accessed November 28, 2014.

Huijboom, N., & Van Den Broek, T. (2011). Open data: An international comparison of strategies. *European Journal of ePractice, 12*(1–13), 4–16.

Hung, S.-Y., Chang, C.-M., & Kuo, S. R. (2013). User acceptance of mobile e-government services: An empirical study. *Government Information Quarterly, 30*(1), 33–44. doi:10.1016/j.giq.2012.07.008.

Hunter, F. (2013). *The civic innovation ecosystem blooms in 2012. What's next?* innovatesf.com/category/open-data/. Accessed November 28, 2014.

IBM. (2015). *Press release* https://www-03.ibm.com/press/us/en/pressrelease/46566.wss. Accessed June 2, 2015.

Immonen, A., Palviainen, M., & Ovaska, E. (2014). Requirements of an open data based business ecosystem. *IEEE Access, 2*, 88–103. doi:10.1109/ACCESS.2014.2302872.

Janssen, K. (2011). The influence of the PSI directive on open government data: An overview of recent developments. *Government Information Quarterly, 28*(4), 446–456. doi:10.1016/j.giq.2011.01.004.

Jetzek, T., Avital, M., & Bjørn-Andersen, N. (2013). Generating value from open government data. In *Thirty Fourth International Conference on Information Systems, Milan* (20 pp).

Johnson, P., & Robinson, P. (2014). Civic Hackathons: Innovation, procurement, or civic engagement? *Review of Policy Research, 4*(31). doi:10.1111/ropr.12074.

Kitchin, R. (2014). The real-time city? Big data and smart urbanism. *GeoJournal, 79*(1), 1–14. doi:10.1007/s10708-013-9516-8.

Kuhn, K. (2011). Open government data and public transportation. *Journal of Public Transportation, 14*(1), 83–97.

Lathrop, D., & Ruma, L. (2010). *Open government: Collaboration, transparency, and participation in practice*. California: O'Reilly Media.

Liu, Y., Li, H., Kostakos, V., Goncalves, J., Hosio, S., & Hu, F. (2014). An empirical investigation of mobile government adoption in rural China: A case study in Zhejiang province. *Government Information Quarterly, 31*(3), 432–442.

Mainka, A., Hartmann, S., Meschede, C., Stock, W. G. (2015a). Mobile application services based upon open urban government data. In *Proceedings of the iConference 2015: Create, Collaborate, Celebrate, Newport Beach, California, USA*, March 24–27, 2015. University of Illinois at Urbana-Champaign: iSchools, IDEALS.

Mainka, A., Hartmann, S., Stock, W. G., & Peters, I. (2015b). Looking for friends and followers: A global investigation of governmental social media use. *Transforming Government: People, Process and Policy, 9*(2), 237–254. doi:10.1108/TG-09-2014-0041.

McDermott, P. (2010). Building open government. *Government Information Quarterly, 27*(4), 401–413. doi:10.1016/j.giq.2010.07.002.

Mergel, I. (2013). Social media adoption and resulting tactics in the U.S. federal government. *Government Information Quarterly, 30*(2), 123–130. doi:10.1016/j.giq.2012.12.004.

Microsoft. (2014). *Costumer storie*. https://customers.microsoft.com/Pages/CustomerStory.aspx?recid=270. Accessed June 2, 2015.

Millard, J. (2010). Government 1.5: Is the bottle half full or half empty? *European Journal of ePractice, 9*, 1–16.

Moro, G. (2005). Citizens` evaluation of public participation. In OECD (Ed.), *Evaluating public participation in policy making* (pp. 109–125). Paris: OECD.

OECD/ITU. (2011). *M-Government. Mobile technologies for responsive governments and connected societies*. Organisation for Economic Co-operation and Development (OECD)/ International Telecommunications Union (ITU), 2011.

OKF. (date unknown). *Open data handbook documentation/open knowledge foundation (OKF)*. Online: opendata-handbook.org/pdf/OpenDataHandbook.pdf.

Reddick, C. G. (2014). Citizens and mobile government adoption: A comparison of activities and uses. *International Journal of Civic Engagement and Social Change, 1*(1), 13–26. doi:10.4018/ijcesc.2014010102.

Scholl, H. J. (2012). Five trends that matter: Challenges to 21st century electronic government. *Information Polity, 17*(3–4), 317–327. doi:10.3233/IP-2012-0280.

Scholl, H. J., & Scholl, M. C. (2014). Smart governance: A roadmap for research and practice. In *Proceedings of iConference 2014* (pp. 163–176).

van der Waal, S., Węcel, K., Ermilov, I., Janev, V., Milošević, U., & Wainwright, M. (2014). Lifting open data portals to the data web. In S. Auer, V. Bryl, & S. Tramp (Eds.), *Linked open data—Creating knowledge out of interlinked data* (pp. 175–195). Berlin: Springer. doi:10.1007/978-3-319-09846-3_9.

Venkatesh, V., Sykes, T. A., & Venkatraman, S. (2014). Understanding e-Government portal use in rural India: Role of demographic and personality characteristics. *Information Systems Journal, 24*(3), 249–269.
Wien.at. (2014). *Nutzungsbedingungen*. https://open.wien.at/site/open-data/nutzungsbedingungen/. Accessed November 28, 2014.

Appendix: List of Hackathons Mentioned in this Chapter

Apps for Amsterdam. (2012). *Second Apps for Amsterdam contest: Smart apps awarded*. www.appsforamsterdam.nl. Accessed November 28, 2014.
Apps for Europe. (n.d.). *About us*. www.appsforeurope.eu/about-us. Accessed November 28, 2014.
Apps for Smart Cities. (2012). *The apps for smart cities manifesto*. Online: www.appsforsmartcities.com/?q=manifesto
Apps4Finland. (November 08, 2012) *Open data innovation contest Apps4Finland boots the clock speed of the finish society*. www.apps4finland.fi/en/. Accessed November 28, 2014.
Barcamp. (2013). *OpenDataHackathon 2011*. www.barcamp.at/OpenDataHackathon_2011. Accessed November 28, 2014.
Cleanweb Worldwide. (2012). *About*. nyc.cleanweb.co/about/. Accessed November 28, 2014.
Digitales Österreich. (n.d.). *Teilnahmebedingungen zum Wettbewerb apps4austria*. www.digitales.oesterreich.gv.at/site/7772/default.aspx. Accessed November 28, 2014.
Studentcompetitions. (2012). *Open stockholm award winners bring stockholm closer towards vision 2030*. studentcompetitions.com/posts/open-stockholm-award-winners-bring-stockholm-closer-towards-vision-2030. Accessed November 28, 2014.

Citizens' Acceptance of U-Life Services in the Ubiquitous City Songdo

Aylin Ilhan, Rena Möhlmann and Wolfgang G. Stock

Abstract As prototypical cities of the twenty-first century, "smart" and "ubiquitous" cities (u-cities) are planned and constructed all over the world. A paradigmatic example of a u-city built from scratch is Songdo in South Korea. U-cities are concerned with the application of the Internet of Things (IoT) in all spaces of the city—including households (with u-life services integrated in the apartments). A crucial aspect of newly built u-cities is the urbanity as it is perceived by the citizens. Our two leading research questions are as follows: Do Songdo's citizens really accept the u-life services as parts of their households (as the centers of their private lives)? Do Songdo's inhabitants recognize this new city as "urban?" The making of Songdo is highly dependent on political decisions and on the activities of the construction and the ICT companies, but not on the people living in Songdo. How is it possible to integrate the people, i.e., the actual and future inhabitants, into planning activities of further development of the u-city? We argue that participatory urbanism is in need of survey tools to collect the citizens' opinions.

Keywords U-life services · U-city · Ubiquitous city · Songdo · South Korea · Smart city

A. Ilhan (✉) · R. Möhlmann · W.G. Stock
Department of Information Science, Heinrich Heine University Düsseldorf, Düsseldorf, Germany
e-mail: aylin.ilhan@hhu.de

R. Möhlmann
e-mail: rena.moehlmann@hhu.de

W.G. Stock
e-mail: stock@phil.hhu.de

M. Foth et al. (eds.), *Citizen's Right to the Digital City*,
DOI 10.1007/978-981-287-919-6_12

1 Introduction

The twenty-first century is characterized by fast-growing information needs. Developers all over the world are designing and constructing so-called "informational", "smart", and "ubiquitous" cities (u-cities). Even the apartments in u-cities are chock-full of information and communication technology (ICT) and of information services (u-services). Can such a u-cityU with its elaborated u-services really satisfy the residents' information needs in his or her private households? And how can the city's administration and the developing companies understand the information needs of their citizens? There are some smart cities and u-cities, which are developed "from scratch," including New Songdo City, which is our case study. How urban is such a planned u-city? Sassen (2012) queries whether smart cities are indeed urban. Is Jacobs (1961, p. 228) right, when she says that "new ideas must use old buildings?" However, not only buildings and ICT are important, people are important, too. As Thackara (2005, p. 187) states "the killer application [of cities] is access to other people. People are what make cities different from other places." "Ubiquitous" means "everywhere," but Kukka et al. (2014, p. 663) add the connotation "for everyone." So we are confronted with the triad of place (the city, its buildings and architecture; Anthopoulos and Fitsilis 2014), technology (ICT, u-services), and people (Foth et al. 2011). Foth et al. (2011) claim that "the main challenge to bring about significant changes in societies is to ensure equal access to technologies and associated literacy skills." How can we ensure that citizens will articulate their ideas and participate in further city development? The simple answer could be, "Ask them!" Till now, the making of Songdo is highly dependent on political decisions and on the activities of the construction and the ICT companies, but not on the people living in Songdo (Halegoua 2011). In order to study the citizens' evaluation of Songdo's u-services and their estimations of the grade of urbanity of a planned new city like Songdo, we designed a questionnaire, two of us (AI, RM) travelled to Korea and asked the people on-site. Our leading research questions are as follows:

- Do Songdo's citizens accept the u-services as parts of their households (as the centers of their private lives)? and
- Do Songdo's inhabitants recognize this new city "from scratch" as urban?

Our chapter is structured into seven paragraphs. After this introduction (1), we go to define some basic concepts as "ubiquitous city," "smart city," and "informational city" (2). Then, we discuss urbanity with emphasis on planned new u-cities (3). In (4), our case study (Songdo) will be described. In the methods part (5), we will introduce two methods of questionnaires which founded our interviews on-site: SERVQUAL and Customer Value Research. The results in Section (6) will present our findings on the acceptance of the u-services in the flats and the interviewees' estimations of Songdo's grade of urbanity. In the discussion (7), we explore options of citizen participation in the further development of u-cities and of u-services.

2 Defining Ubiquitous City, Smart City, and Informational City

Jang and Suh (2010, p. 263) state that a ubiquitous or "u-city is a twenty-first century futurist city." According to Shin and Kim (2012, para. 2.1.2), the availability of "wireless high-speed networks and advanced information services regardless of location through a ubiquitous computing network" is the main goal of a u-city. Lee et al. (2008) see the ubiquitous computing as an important key factor for the ICT, which not only has an effect on society, but also on urban planning and city competitiveness. ICT supports the decision and action of residents. According to De Almeida (2009), the availability of ubiquitous computing will be the same as it is for energy today, being accessible anywhere and anytime. Ubiquitous computing is a technology embedded in our everyday life (Anttiroiko 2013). "The idea of integrating computers seamlessly into the world" is according to Weiser (1991, p. 94) the objective of ubiquitous computing. Ubiquitous infrastructure includes a spectrum of aspects of the Internet of Things (IoT), which include sensors and radio-frequency identification (RFID). According to Haller et al. (2009, p. 15), IoT has the characteristic that "physical objects are seamlessly integrated into information networks."

What does the term "smart city" mean? We can find two different concepts of the smartness of cities, one in a narrow sense of the concept and the other in a much broader sense (Fietkiewicz and Stock 2015). According to Chourabi et al. (2012, p. 2289), it is possible "to conceptualize a smart city […] as an icon of a sustainable and livable city." In line with Hall et al. (2000, p. 1), "the vision of 'Smart Cities' is the urban center of the future, made safe [and] secure environmentally green." ICT takes an important role to fulfill the aims of a smart city. This narrow concept of "smartness" is strongly linked to natural resources and energy, transport and mobility, buildings, and living conditions, in short, to the green, sustainable, and livable city.

Giffinger et al. (2007) define smart cities far broader by an enumeration of essential "characteristics," i.e., smart economy, smart people, smart governance, smart mobility, smart environment, and smart living. Such a broad view of smart cities is anticipated by Castells as early as 1989. He calls such prototypical cities of the network society "informational cities." In informational cities, two spaces exist side by side: geographical space ("space of places") and the space of information, money, and power streams ("space of flows"). In informational cities (or smart cities in the broader sense), the space of flows outperforms the space of places. The concept of informational city (Stock 2011) includes the narrower concept of the smart (green) city.

3 Urbanity of Planned New U-Cities

Urbanity is a central characteristic of every city. Van Diepen and Musterd (2009, p. 333) identify urban life as follows: "urban residence provides households with, among other things, ample availability of facilities and services and the opportunity to minimize the coupling constraints on performing various daily tasks and recurring activities." Van Diepen and Musterd (2009, p. 332) not only look at the urban households, but they also try to find the "linkages between urbanity and urban residents." They also state that "usually, urbanity refers to centrally located city sites where urban values such as solidarity, tolerance and the enjoyment of differences can exist." It is in "the pragmatic, daily practices of a city's pubs and restaurants, streets and shops, that 'pictures of a functioning urbanity' come into being" (Van Diepen and Musterd 2009, p. 335).

Sassen (Interview with Nova 2011) argues that technology can be urbanized but "that the model of 'intelligent cities' as propounded by technologists, with the TelePresence efforts of Cisco Systems a key ingredient, misses this opportunity to urbanize the technologies they mobilize." Furthermore, she issues the statement that "the biggest challenge for intelligent cities like Songdo will be to design their services in a way that the technology will be of use for the residents and not vice versa" (Sassen 2012, p. 3 translated). Kim (2010, p. 16) mentions that the "wireless, context-aware, computer-driven information and communication technologies" are used to present Songdo as fulfilling "nearly all aspects of urban life." Sassen also states that "spaces like Songdo deurbanize a city." For Sassen, Songdo is not really an urban space, but a platform where everything is functioning, everything is efficient, and all concerns of a high specialized businessman are discerned—whether at the working space, at home, or on the way in between. The intelligent city dabbles in being a perfect, closed system and is thus the opposite of urban and alterable—"and that is going to end its life early" (Sassen 2012, p. 3 translated). Regarding the aspect of planning a city and to be urban, Lees (2010, p. 2302) argues, too, that "on the surface it seems bizarre that the planning profession, developers, and policy makers talk about urbanity as if it were something definable and indeed attainable." Lees further discusses the aspect of "a planned approach to urbanity" and brings up the question of "how are they [the planners] planning urbanity and what are the results of this?" (Lees 2010, p. 2303). The marketing material has to be viewed carefully though as things are often pictured almost utopic, "far removed from real life" (Ward 1998, p. 239).

4 Case Study: New Songdo City

New Songdo City is situated at a vacant lot of about 1,415 acres approximately 40 miles away from Seoul in South Korea. "Korea is the first country to envision and realize u-City" (Hwang 2009, p. 367). The reclaimed city is part of Incheon

(Lee et al. 2008). Prior to reclaiming the land, there had only been water in the area Songdo was built. Songdo is a ubiquitous and smart city "from scratch." Such a city is "planned in social and physically 'virgin land' (that is, with no former residents, buildings or infrastructures), in policy-protected arenas (with loose, experimentation-oriented and flexible regulations), aiming to attract new residents and companies to be simultaneously users and developers of new smart-IT solutions" (Carvalho 2015, p. 48). It is not unimportant to know that Songdo is "a self-declared ubiquitous eco-city or u-eco-city, as it strives to contribute to the making of a 'Green Korea'" (Shwayri 2013, pp. 39–40). The makers of Songdo are both the Korean policymakers and the involved companies. For Kim (2013, p. 329), Songdo is "designed by top-down decisions." Carvalho (2015, p. 50) states, "u-solutions have been largely pushed from corporations to residents, in a rather inflexible fashion." We were not able to detect any larger citizens' participation in making Songdo.

It is estimated that Songdo has accommodation for 65,000 residents and 300,000 workers in 2014 (Lee and Oh 2008). Songdo is intended to be "the principal business hub in northeast Asia" (Segel 2005, p. 1). It benefits from "strategic location, advanced infrastructure, and business-friendly environment" (Segel 2005, p. 1). The artificial structure of Songdo is due to the American developer Gale International and the Korean construction firm POSCO E&C (Kim 2010). Gale International has been involved in the project New Songdo City since 2001. "They planned an urban complex that was ambitious, futuristic, and utopian" (Kuecker 2013, p. 5). Cisco is involved as well. As the developer of the u-services in Songdo, it is responsible for the ICT (Halpern et al. 2013). Songdo has several facilities, buildings, and leisure time opportunities such as the central park. With the New York City's Central Park in mind, a park, even named "central park" as well, has been built. Moreover, the system of the canals in Venice is reflected at the canal walk in New Songdo City (O'Connell 2005). In addition, Songdo has international schools, a convention center, museum, and golf clubs. In all these components which Songdo features, the postmodern architecture is visible (Kim 2010). Songdo is connected with services of the u-life "where wireless, context-aware, computer-driven information and communication technologies imbue nearly all aspect of urban life" (Kim 2013, p. 16). For Kim (2013), Songdo represents a paradigmatic example of an informational city in the sense of Castells (1989). "Patterned on the special economic zone (SEZ) and given a territorial exemption from the state's regulatory power, the international free economic zone of Songdo is South Korea's regulatory power, the international free economic zone of Songdo is South Korea's response to what Manuel Castells calls a globalized 'space of flow'" (Kim 2013, p. 331).

5 Methods

In the literature, we were able to identify a "citizen-centric typology for smart city services" (Lee and Lee 2014, p. S93). But this typology fails for the dimension of households. Lee and Lee (2014, p. S104) only mention in their type "miscellaneous" home management services, outside linkage services, and housing complex management services. In contrast, we will concentrate on households and on the residents. To get quantitative as well as qualitative answers for our research questions, different methods were used. With regard to the theoretical background and the specifics of Songdo, the first aim was to find out what u-services were already implemented. The most important question is as follows: How satisfied are the residents with those u-services? On the urban level, we want to study whether Songdo is a really urban city in the eyes of its residents.

For our project, it was important to have interviews with residents or other people who potentially know enough about Songdo. The challenge was to find suitable interview partners first. It is important that on the one hand, they know their city and on the other hand, they are willing to talk to foreign scientists. Who can be a suitable interview partner? In this case study, a suitable interview partner could be each resident of Songdo and also people who visited Songdo, developers, or people who did research in Songdo. We had contacted professors, residents via Facebook groups, researchers, and staff of Gale and Cisco before traveling to Songdo.

We interviewed 21 people (there of 14 students) and 2 representatives of Gale and Cisco. The two parts of the ServQual-like quantitative questionnaire as well as the qualitative interviews were conducted at the same time. The interviews were of different length, with the time frame being between 1 and 2 h.

The SERVQUAL method was used to collect expectation and experience values of residents in Songdo on the u-city's u-services. For this purpose, a SERVQUAL-like questionnaire (Parasuraman et al. 1988) was developed. It contained 21 questions with a seven-point Likert scale (for details of the questionnaire see Ilhan et al. 2015). The SERVQUAL part of the questionnaire deals with the expectation value, the experience value, and the gap value (this is experience value minus expectation value) (Stock and Stock 2013, p. 482).

To guarantee qualitative results, the in-depth interview part was added and the questionnaire was used as a guideline for interviews. The interviews allowed the interviewee to add comments and state problems that occur with the systems mentioned in the SERVQUAL part. Furthermore, four non-SERVQUAL interview questions were included to learn about how the residents view the aspects of urbanity.

Based on McKnight (2006), the SERVQUAL method was expanded to the Customer Value Research. The SERVQUAL method concentrates only on the residents, but Customer Value Research focuses on developers as well. This method can be used to provide an insight to the "irritation" between the customer experience values and the expectation values of the developers. The irritation uncovers the misunderstanding between what the developers think their customers want and what the customers actually want and need. The irritation value is calculated by subtracting the modified expectation value of the developers from the experience value of the residents. We are talking about modified expectation of the developer because developers should have to ask themselves for each question "What do I expect that my customer said?"

For the Likert scale gives ordinal data, we worked with the median as mean value for all gap analyses and the quantitative data from the Customer Value Research. Due to the very small size of our data, the statistical results should be interpreted with great caution. But the qualitative results from the in-depth interviews with the residents in Songdo (N = 21) support the (weak) quantitative data.

6 U-Life Services in the Eyes of Songdo's Residents

Our research shows that some of the u-life services are already in use and some u-services that are introduced in advertising videos are not yet ready. It has to be taken into account though that Cisco "only [joined the project] in 2009, five years after buildings started rising above the landfill" (Strickland 2011, p. 12). They are working on equipping existing buildings with their technology. Cisco's director Munish Khetrapal has an aim concerning the u-life services, "We are trying to make this equipment part of the standard system, […] when you turn on the tap you have water to drink, and when you turn on the TelePresence you have a conference" (Strickland 2011, p. 12). So, partly the u-services for the households can be seen as "smart sustainable" services, but others are (still) "branding hoaxes" (Yigitcanlar and Lee 2014).

The highlight for most of Songdo's residents is the trash system (in Table 1: no. 1). We saw two different kinds of composition of the trash system. In one kind of apartments, every corridor has its own waste chamber that can be opened with the residents' card. In the second apartment type, trash systems were not in the corridors. The residents of these apartments have to dispose of the trash in the forecourt where the trash system was furnished. The interview partners presented

Table 1 Realized exemplary u-services in New Songdo City (2014)

U-services	Realized?
1. Central trash system	Yes
2. Master panel	Yes
3. Audio/video (cameras)	Yes
4. Temperature control	Yes
5. Smart meter	Yes
6. Lighting control	Yes
7. Card access	Yes
8. WiFi (flat/city)	Yes
9. Remote control (via phone)	No
10. Intelligent fridge	No
11. RFID chips/sensors	No
12. Systems (water, power, etc.) connected	No
13. Smart health	No
14. Media poles	No

us the master panel (2) that is installed in every apartment. The master panel has several functions and integrates many u-services. At the front door, at the playgrounds in the apartment complex and in the lobby, cameras are installed (3). The residents can switch between the different camera views via the master panel. This option is not only available on the master panel in the living room but also in the bathroom. So in case that someone is knocking on your door, you can see who it is. You control the room temperature and the air quality (4) via the master panel. The master panel gives an overview about the usage of power. Here, it acts as a smart meter (5). Furthermore, it can be used as a telephone, to control the lighting (6), or as a memo board. For the foreigners, there is the option to obtain a master panel in English. With the residents' cards (7), all doors (building, flat, trash system, etc.) can be opened. Free wireless Internet was not available at the time we were in Songdo. But the alternative exists to login into most of the free wireless services of the restaurants, bars, shops, and also at the central park in some places. Of course, commercial Internet access (8) was realized in Songdo.

None of the other u-life services (9–14) were available during our stay in Songdo, but there were prototypes of u-life services in the showroom of Cisco that we had seen in different videos. We got the chance to meet interview partner 2 who was chosen as a test person for a product of Cisco. The test was about the usage of a product that resembles a television receiver. It enables the communication between residents in different apartments like Skype.

Figure 1 represents an overview on the acceptance values of u-life services in Songdo. The first question was about audio and video in every room for easier communication. The evaluation shows that the customers do not that much expect audio and video in an apartment in a ubiquitous city (median: 2.5). The median experience is 3, leading to a gap of +0.5. The interviewed residents expect lighting control in a u-city with a value of 4 and notice the system as very useful (6). The gap value of +2 is a signal of high acceptance of this service. An optimal evaluation goes to the service of card access as both values have a median of 7. Similar successful is the service of the central trash system (expectation: 6, experience: 6.5, gap: +0.5). Smart meters and the temperature and air quality control exhibit high expectation scores (6.5 and 7), but a lower experience value leading to a gap value of −1.5 for smart meters and of −1 for temperature control. Here is room left for improvements. WiFi in flats is no problem (experience: 7; gap: +3), but WiFi in city areas is wanted (expectation: 6) but problematic (experience: 4). This gap (−2) calls for perfecting open WiFi city-wide.

Figure 2 shows the results on the residents' acceptance of some urban aspects in a u-city built "from scratch." The residents expect heavily (7) green areas. Indeed, the median of the experience (6) shows that Songdo does not fulfill residents' expectation to the full extend. Songdo offers an international university. During the interviews and conversation with the students, we found out that the students want to make a semester abroad in countries such as America. The evaluation shows that the experience is more pleased (7) than expected (6). With regard to leisure activities, the score shows that the interviewed persons are very disaffected with cultural facilities (−2). Songdo offers a lot of pubs, restaurants, and

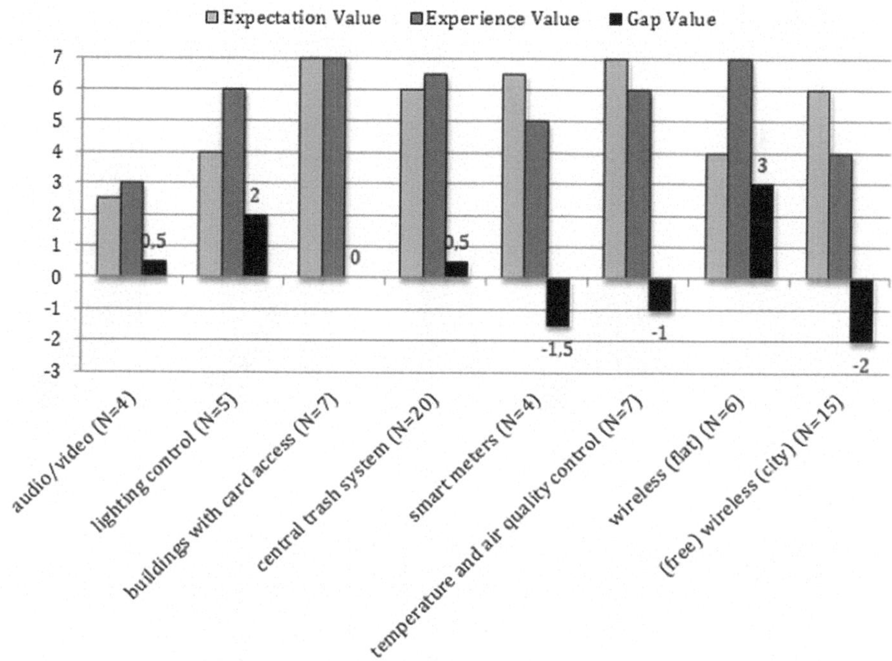

Fig. 1 Acceptance of the u-services by the residents (median values)

bars, but the gap value (−2) shows that the citizens want even more. The prices for rents or for buying a flat in Songdo seem to be okay for the residents.

Figures 3 and 4 exhibit the amount of irritation between the developers (Gale and Cisco) and the residents regarding u-life services (Fig. 3) and urbanity (Fig. 4). Concerning the WiFi in flats and the card access, the irritation value

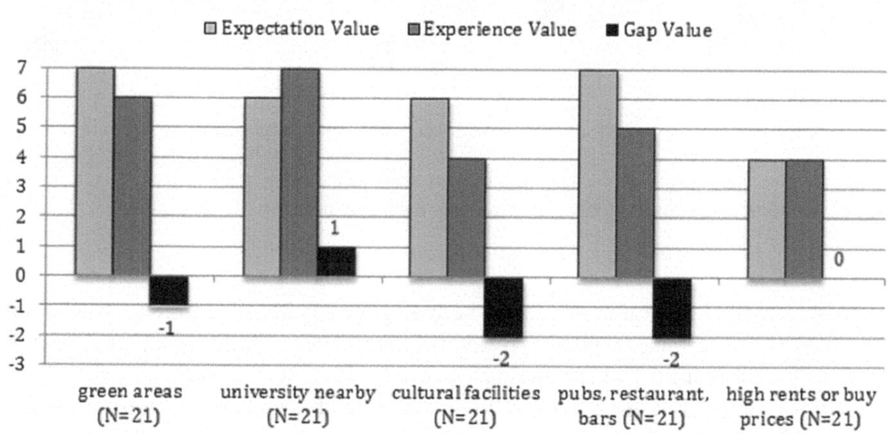

Fig. 2 Acceptance of some aspects of urbanity in a u-city "from scratch" (median values)

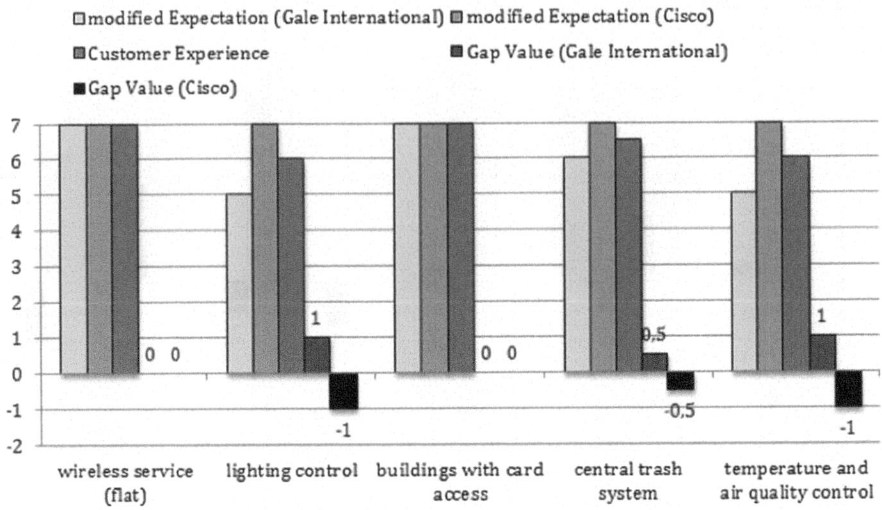

Fig. 3 Irritation between developers and residents regarding u-services (median values)

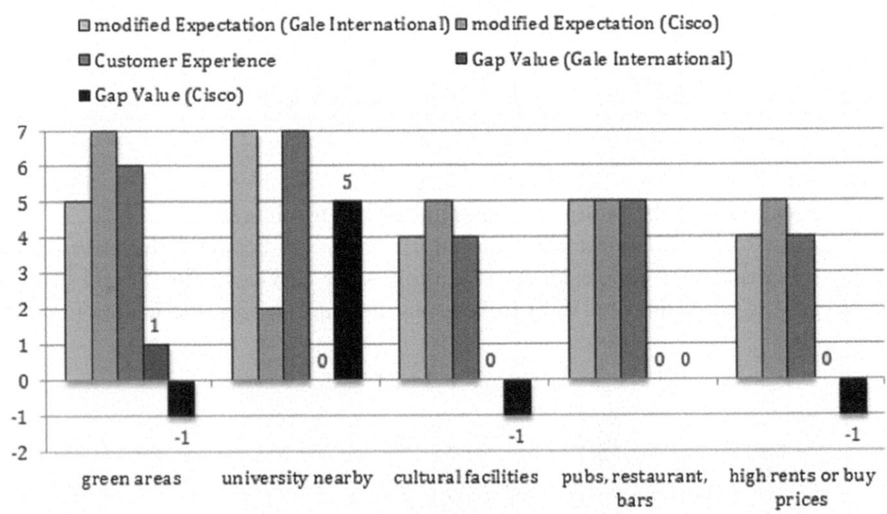

Fig. 4 Irritation between developers and residents regarding urbanity (median values)

equals 0, which is the optimal result. There are small negative differences between the estimations of Cisco and the residents' experiences: lighting control (irritation: −1), temperature and air quality control (−1), and the trash system (−0.5). In contrast, Gale estimates more cautious, which leads to positive irritation values. Concerning urbanity, Cisco fails again in the aspects of green areas, cultural facilities, and buying prices (each −1). Additionally, Cisco underestimates the role of a university nearby. While for the residents, their experiences with the local university are very good (7), the representative from Cisco estimated the user experience with 2, leading to a gap of +5. Also in regard to the analyzed aspects of urbanity, Gale estimates the users' experiences quite well.

Regarding what Sassen said about urbanity the residents have different opinions. The question "Do you really live in an urban city?" shows distinct opinions. 27 % of the interviewees (N = 15) say "no, I do not really live in an urban city" and also 27 % say "not yet." But 46 % are of the mind that they live in an urban city. For example, Interviewed Person 3 answered this question, "no (40–50 % occupied). People mostly just sleep here."

Interview Person 1 told us, "Not yet, but I think in 5 or 7 years it will be a new city." One further answer was "Yes, Songdo is definitely urban but relatively under-populated for the number of buildings and the infrastructure. For me, perhaps as I am not Korean and do not plan to live here forever, it lacks a soul" (Interview Person 4). Finally, Interview Person 7 said, "As Songdo is not finished yet it has more a village feeling."

It is also interesting to see how the developers answered when being asked, "Do you really live in an urban city?" While Jonathan Thorpe (Interview Person 5) of Gale International tends to agree with the theory of Sassen "that you only have one sort of people and not layers," Cisco answers on the contrary, "Songdo has every component that a city required such as universities, schools, parks, companies, hospitals and so on" (Interview Person 6).

It is important that a lot of aspects of urbanity interact with each other. We could affirm these impressions of the residents. Most of the time, streets were unpeopled. During the interviews, it became apparent that the city does need time to fulfill its potential. Most of our interviewed persons do, however, believe in the future of Songdo as an urban place.

The residents were also asked whether they think that cities like Songdo should be built in other parts of the world. From 23 interviewed person, only one interviewee answered with "no" and 13 % with "not yet." In contrast, 83 % said "yes, cities like Songdo should be built." Our interviewees explained their answers with opinions such as "yes, it improves the quality of life of population. And it changes the way of people lived in a city and in community" (Interview Person 1).

7 Discussion

Songdo is a planned new u-city "from scratch." It is at an early stage concerning urbanity and ubiquitous services in the households. The analysis shows that some u-services are installed (as the master panel, the trash system, and the digital access card) and some others are planned, but not realized yet. At the time, the city is not really alive. Songdo is missing a lot of people, who do not only sleep or work in Songdo but also live there. We agree with the opinion that the city needs time to develop itself, but an urban infrastructure is in place. Moreover, according

Fig. 5 Control room in G Tower, Songdo. *Photograph* A. Ilhan

to Shwayri (2013), it is true that some re-orientations, caused by different facts and conflicts, happened while building Songdo. There are apartment blocks, where you can bring your child to a nursery, or bring your dogs to the dog café or get a massage. These shops are located on single floors of the apartment buildings. For the future work, it will be important to check the state of affairs, if more u-services and more shops, restaurants, etc., are really implemented, as they are supposed to and what happened to the construction infrastructure.

Finally, we have to note some open questions. In the course of the practical application of u-life services in the households of Songdo, huge amounts of data become accumulated. For example, data from smart meters (e.g., energy consumption) become observed real time in the buildings' control rooms (Fig. 5) and are stored in big archives. All actions with u-services (videos, telephone calls via the master panel, every usage of the card, etc.) can be objects of data storage. How is it safeguarded that there will be no misuse of the data?

Songdo is an "innovation laboratory" (Carvalho 2012, p. 198) for the development of smart and ubiquitous cities. The construction of Songdo was and is driven by policymakers and by the leading companies, most notable Gale International and Cisco. Hollands (2015, p. 61) states that the development of cities like Songdo "has left little room for ordinary people to participate in the smart city." How is it possible to integrate the people in city making in order to constitute participatory urbanism? Kukka et al. (2014a) discuss methods as fictional storytelling, use of mock-up devices, technology diaries, and interviews to engage a city's community for participating in future city making. We agree with Kukka et al. (2014)

and emphasize the application of interviews and surveys to reach the residents. SERVQUAL-like gap analyses uncover the differences between people's expectations of (any) service and their experience with realized ones. Customer Value Research gives hints at discrepancies ("irritations") between the residents' experiences and the developers' views. Our practical advice for policymakers and companies to reach and to engage the citizens is very simple: "Ask them!" And: "Discuss the results of interviews and surveys with them!" In this way, questionnaires and in-depth interviews with residents are tools for the first steps in fostering citizen participation in city making.

References

Anthopoulos, L., & Fitsilis, P. (2014). Exploring architectural and organizational features in smart cities. In *16th International Conference on Advanced Communication Technology* (pp. 190–195). Washington, DC: IEEE.

Anttiroiko, A. V. (2013). U-cities reshaping our future: reflections on ubiquitous infrastructure as an enabler of smart urban development. *AI and Society, 28*(4), 491–507. doi:10.1007/s00146-013-0443-5.

Carvalho, L. (2012). Urban competitiveness, u-city strategies and the development of technological niches in Songdo, South Korea. In M. Bulu (Ed.), *City Competitiveness and Improving Urban Subsystems: Technologies and Applications* (pp 197–216). Hershey, PA: IGI Global. doi:10.4018/978-1-61350-174-0.ch010.

Carvalho, L. (2015). Smart cities from scratch? A socio-technical perspective. *Cambridge Journal of Regions, Economy and Society, 8*(1), 43–60. doi:10.1093/cjres/rsu010.

Castells, M. (1989). *The informational city. Information technology, economic restructuring, and the urban-regional process*. Oxford, UK: Blackwell.

Chourabi, H., Nam, T., Walker, S., Gil-Garcia, J. R., Mellouli, S., Nahon, K., Pardo, T. A., & Scholl, H. J. (2012). Understanding smart cities: An integrative framework. In *Proceedings of the 45th Hawaii International Conference on System Sciences* (pp. 2289-2297). Washington, DC: IEEE Computer Society.

De Almeida, M. D. C. P. (2009). *Ubiquitous computing and natural interfaces for environmental information*. Thesis: Universidade Nova de Lisboa.

Fietkiewicz, K. J., & Stock, W. G. (2015). How "smart" are Japanese cities? An empirical investigation of infrastructures and governmental programs in Tokyo, Yokohama, Osaka and Kyoto. In *Proceedings of the 48th Hawaii International Conference on System Sciences* (pp. 2345–2354). Washington, DC: IEEE.

Foth, M., Choi, J. H. J., & Satchell, C. (2011) Urban informatics. In *Proceedings of the ACM 2011 Conference on Computer Supported Work* (pp. 1–8). New York, NY: ACM. doi:10.1145/1958824.1958826.

Giffinger, R., Fertner, C., Kramar, H., Kalasek, R., Pichler-Milanovic, N., & Meijers, E. (2007). *Smart cities—ranking of European medium-sized cities*. Vienna, Austria: Centre of Regional Science.

Halegoua, G. (2011). The policy and export of ubiquitous place: Investigating South Korean u-cities. In M. Foth, L. Forlano, C. Satchell, & M. Gibbs (Eds.), *From social butterfly to engaged citizen: Urban informatics, social media, ubiquitous computing, and mobile technology to support citizen engagement* (pp. 315–334). Cambridge, MA: MIT.

Hall, R. E., Bowerman, B., Braverman, J., Taylor, J., Todosow, H., & von Wimmersperg, U. (2000). The vision of a smart city. In *2nd International Life Extension Technology Workshop*. Paris, France.

Haller, S., Karnouskos, S., & Schroth, C. (2009). The internet of things in an enterprise context. *Lecture Notes in Computer Science, 5468*, 14–28.

Halpern, O., LeCavalier, J., Calviloo, N., & Pietsch, W. (2013). Test-bed urbanism. *Public Culture, 25*(2), 273–306. doi:10.1215/08992363-2020602.

Hollands, R. G. (2015). Critical interventions into the corporate smart city. *Cambridge Journal of Regions, Economy and Society, 8*(1), 61–77. doi:10.1093/cjres/rsu011.

Hwang JS (2009) u-City: The next paradigm of urban development. In: Foth M (ed.) *Handbook of research on urban informatics: The practice and promise of the real-time city* (pp. 367–378). IGI Global, Hershey, PA. doi:10.4018/978-1-60566-152-0.ch025.

Iihan, A., Möhlmann, R., & Stock, W. G. (2015) Customer value research and servqual surveys as methods for information need analysis. In F. Pehar, C. Schlögl & C. Wolff (Eds.), *Re-inventing information science in the networked society. Proceedings of the 14th International Symposium on Information Science* (pp. 457–468). Hülsbusch, Glückstadt, Germany.

Jacobs, J. (1961). *The death and life of great American cities*. New York, NY: Random House.

Jang, M., & Suh, S. T. (2010). U-city: New trends of urban planning in Korea based on pervasive and ubiquitous geotechnology and geoinformation. *Lecture Notes in Computer Science, 6016*, 262–270. doi:10.1007/978-3-642-12156-2_20.

Kim, C. (2010). Place promotion and symbolic characterization of New Songdo City. *South Korea Cities, 27*(1), 13–19. doi:10.1016/j.cities.2009.11.013.

Kim, J. I. (2013). Making cities global: The new city development of Songdo, Yujiapu and Lingang. *Planning Perspectives, 29*(3), 329–356. doi:10.1080/02665433.2013.824370.

Kuecker, G. D (2013). *Building the bridge to the future: New Songdo City from a critical urbanism perspective*. Centre of Korean Studies Workshop: University of London.

Kukka, H., Luusua, A., Ylipulli, J., Suopajärvi, T., Kostakos, V., & Ojala, T. (2014a). From cyberpunk to calm urban computing: Exploring the role of technology in the future cityscape. *Technological Forecasting and Social Change, 84*, 29–42. doi:10.1016/j.techfore.2013.07.015.

Kukka, H., Ylipulli, J., Luusua, A., & Dey, A. K. (2014). Urban computing in theory and practice: Towards a transdisciplinary approach. In *Proceedings of the 8th Nordic Conference on Human-Computer Interaction: Fun, Fast, Foundational*. New York, NY: ACM (pp 658–667). doi:10.1145/2639189.2639250.

Lee, S. H., Han, J. H., Leem, Y. T., & Yigitcanlar, T. (2008). Towards ubiquitous city: Concept, planning, and experiences in the Republic of Korea. In T. Yigitcanlar, K. Vellbeyoglue & S. Baum (Eds.), *Knowledge-based urban development: Planning and applications in the information era* (pp 149–169). Hershey, PA: IGI Global. doi:10.4018/978-1-59904-720-1.ch009.

Lee, J., & Lee, H. (2014). Developing and validating a citizen-centric typology for smart city services. *Government Information Quarterly, 31*, S93–S105. doi:10.1016/j.giq.2014.01.010.

Lee, J., & Oh, J. (2008). *New Songdo City and the value of flexibility: A case study of implementation and analysis of a mega-scale project*. Master Thesis, Massachusetts Institute of Technology.

Lees, L. (2010). Commentary. *Environment and Planning, 42*(10), 2302–2308. doi:10.1068/a43360.

McKnight, S. (2006). Customers value research. In T. K. Flaten (Ed.), *Management, marketing and promotion of library services based on statistics, analyses and evaluation* (pp. 206–216). Munich, Germany: De Gruyter Saur.

Nova, N. (6 July 2011). *An interview with Saskia Sassen about "Smart Cities"*. http://www.nicolasnova.net/pasta-and-vinegar/2011/07/06/an-interview-with-saskia-sassen-about-smart-cities.

O'Connell, P. L. (2005). Korea's high-tech utopia, where everything is observed. *The New York Times, October 5*, 2005.

Parasuraman, A., Zeithaml, V. A., & Berry, L. L. (1988). SERVQUAL: A multiple-item scale for measuring consumer perceptions of service quality. *Journal of Retailing, 64*(1), 12–40.

Sassen, S. (2012). *Die Global City ist ein brutaler Ort*. WOZ no. 25 (June 21, 2012), 15–17.

Segel, A. I. (2005). *New Songdo City*. Boston, MA: Harvard Business School (9-206-019).

Shin, D. H., & Kim, T. (2012). Enabling the smart city: The progress of u-city in Korea. In *Proceedings of the 6th International Conference on Ubiquitous Information Management and Communication*. New York, NY: ACM, art. 105. doi:10.1145/2184751.2184872.

Shwayri, S. T. (2013). A model Korean ubiquitous eco-city? The politics of making Songdo. *Journal of Urban Technology*, *20*(1),39–55. doi:10.1080/10630732.2012.735409.

Stock, W. G. (2011). Informational cities: Analysis and construction of cities in the knowledge society. *Journal of the American Society for Information Science and Technology, 62*(5), 963–986. doi:10.1002/asi.21506.

Stock, W. G., & Stock, M. (2013). *Handbook of information science*. Berlin, Germany, Boston, MA: De Gruyter Saur.

Strickland, E. (2011). Cisco bets on South Korean smart city. *IEEE Spectrum, 48*(8), 11–12. doi: 10.1109/MSPEC.2011.5960147.

Thackara, J. (2005). The post-spectacular city—and how to design it. In S. Franke & E. Verhagen (Eds.), *Creativity and the City: How the creative economy changes the city* (pp. 184–191). Rotterdam, The Netherlands: NAi Publisher.

Van Diepen, A. M. L., & Musterd, S. (2009). Lifestyle and the city: Connecting daily life to urbanity. *Journal of Housing and the Built Environment, 24*(3), 331–345. doi:10.1007/s10901-009-9150-4.

Ward, S. V. (1998). *Selling places: The marketing and promotion of towns and cities 1850–2000*. London, UK: Routledge.

Weiser, M. (1991). The computer for the 21st century. *Scientific American, 265*(3), 94–104.

Yigitcanlar, T., & Lee, S. H. (2014). Korean ubiquitous-eco-city: A smart-sustainable urban form or a branding hoax? *Technological Forecasting and Social Change, 89*, 100–114. doi:10.1016/j.techfore.2013.08.034.

A Community Architecture Framework for Smart Cities

Tim McGinley and Keiichi Nakata

Abstract The design of smart cities, like urban planning, is a wicked problem. Cities have diverse stakeholders with complex interrelationships and conflicting interests which means that it can be difficult to gain a representative consensus on city design. Typically, few of a city's stakeholders are involved in smart city design. This means that the needs of some stakeholders can be underrepresented. This non-participatory approach is in opposition to what urban planning suggests which leads to sustainable communities. This chapter identifies a participatory approach to the development of an information system to support smart city design. In businesses, enterprise architecture (EA) frameworks are used to link their needs to their information systems. EA frameworks link the operational goals of an enterprise to, among other things, its information systems. Current approaches to smart cities are framed by corporate narratives. This could mean that if the smart city design tools are developed in a standard EA approach, the needs of those that procure and provide the service will be considered above others, for example, the interests of citizens. Therefore, this research proposes a 'community' architecture framework (CAF) and development methodology that can support diverse stakeholder requirements and perspectives. Smart cities are not only wicked problems, but they also represent large spatial, temporal and

T. McGinley (✉)
University of South Australia, Adelaide, Australia
e-mail: Tim.mcginley@unisa.edu.au

K. Nakata
Henley Business School, University of Reading, Reading, UK
e-mail: k.nakata@henley.ac.uk

M. Foth et al. (eds.), *Citizen's Right to the Digital City*,
DOI 10.1007/978-981-287-919-6_13

populational scales. In the light of this, the CAF is tested on a smaller scale in a 'real' city planning context in Chelmsford, UK. This chapter then discusses how this could be used to support the participation of diverse stakeholders in smart city planning.

Keywords Smart cities · Wicked problems · Participatory design · Enterprise architecture · Community architecture framework · Public participation · GIS geographical information systems

1 Introduction

The corporate narrative of smart cities argues that through the procurement of IT services, we can achieve more efficient and intelligent smart cities (Söderström et al. 2014). Söderström et al. (2014) argue that, based on Hollands (2008), these narratives oversimplify the challenges of cities by framing the city as a series of quantifiable challenges, such as pothole location identification and broadband coverage. It is clear that our cities could be more efficient. We could also make better use of existing data and knowledge to understand how our cities work and ultimately build smart[er] cities. Chourabi et al. (2012) state that 'making a city smart is emerging as a strategy to mitigate the problems generated by urban population growth and rapid urbanisation'. From the governmental and consultant perspective, smart cities have been promoted as a technological panacea to the challenges cities face. However, like other forms of urban planning, smart cities are represented by a number of different narratives which reflect the interests of some stakeholders over others.

It is therefore essential that the citizen's voice is represented in the planning of smart cities. However, smart city projects also provide the potential for local government decisions to be outsourced by the back door to the IT systems of private corporations. This is because beyond the initial commission, the maintenance and operation of smart cities offers IT consultancies significant business opportunities that may be in conflict with the needs of the citizens. An alternative narrative for smart cities such as those proposed by Giffinger et al. (2007) is that we could use IT services (digital tools) to democratise the planning of smart cities and consider the health and the improvement of citizen's everyday lives (Anthopoulos and Fitsilis 2010). Söderström et al. (2014) proposed the following definition: *'Smart cities involve the creation of new relations between technology and society'*. It is therefore important to assess how the relations are defined and by whom, as these relations and the technology they employ will ultimately have an effect on cities and their citizens.

Chourabi et al. (2012) describe the post-industrial city with a biological metaphor of an organic system that includes a nervous system, brain and sensory organs. However, we know that embedded legacy systems in cities mean that it would be very difficult to make a whole city run on a single unified system, and in reality, what is required is an orchestration of different systems at different stages of maturity, which includes both new and legacy systems. Therefore, beyond the consultancy narrative, any mature smart city approach might look more like an organism that resembles a technological Frankenstein's monster than the sleek solutions described by corporate narratives. One perspective of smart cities is that they present the opportunity to 'outsource' urban management that would have traditionally been managed by municipalities, to IT companies to run cities on our behalf. The corporate narrative could also mean that there would be new knowledge silos between the cities and their systems depending on the openness of the IT service provider's smart city systems. This could result in a smart city 'monster' that does not know its arms from its elbows.

The corporate narrative of smart cities is an oversimplification; therefore, for a non-simplified narrative, we should look at the urban planning literature. Urban planning has been described as a 'wicked problem' (Rittel and Webber 1973). Every wicked problem is essentially unique; can be seen as a symptom of another problem, and has numerous explanations (Rittel and Webber 1973). Smart cities are a wicked problem. Wicked problems have diverse stakeholders (Weber and Khademian 2008), which makes it necessary to have all stakeholders involved in the planning process (Maquil et al. 2007). The diversity of the stakeholders causes them to adopt different perspectives on the origins, severity and existence of the problems their community faces, which leads to alternative understandings of the best strategies and solutions to address the problems (Rittel and Webber 1973) based on alternative problem narratives (Australian Public Services Commission 2007). The planning conflict triangle (Campbell 1996) represents the three connected conflicts of property, development and resources in planning. The conflict triangle shows that the stakeholders have different perspectives on the challenges facing the city which can lead to a conflict of interests between the stakeholders (Campbell 1996). It would therefore seem that we need multidimensional narratives to represent the perspective of each stakeholder in smart cities. This is needed in order to avoid the needs of stakeholders that procure and provide the 'smart city' service being considered above, for example, the interests of citizens.

2 Participatory Urban Design

Arnstein's ladder of participation (Arnstein 1969) can be used to represent the power of citizens in the urban planning process, from the bottom rung of the ladder, manipulation, to the top rung of the ladder, citizen control. A non-participatory urban planning approach would be towards the bottom of the ladder. This could manifest itself in a way that a local government agency or council could employ designs that it believes are in the best interest of its people, without properly consulting them on what a smart city could mean for them. Smart cities are often planned from a top-down approach (Söderström et al. 2014), leaving little opportunity for citizens to engage in their design. The integration of non-experts into the planning process is a move beyond the traditional advocacy role of urban planners. Participatory alternatives to the corporate narrative of smart cities require an approach to support citizen participation in the planning of smart cities. Previous approaches to supporting citizen participation can be broadly divided into offline and online support. De Cindio et al. (2007) describe three stages of a participatory process that include selection of participants; issue framing; and choice of the participatory modality. De Cindio et al. (2007) model a Deliberative Community Network (DCN) architecture to support three different spaces: the community, informational and the deliberative space.

This chapter explores supporting online participation in smart city planning by creating a deliberative space for citizens and other stakeholders. Online participatory planning systems were originally discussed in the mid-1990s (Sieber 2006) with empirical analysis and presentation of technical system architecture occurring in the early 2000s (Peng 2001). Participatory online technologies can support engagement that is 'more readily accepted and relevant to communities' (Rinner and Bird 2009). An example of these systems is public participation geographical information systems (PPGIS). However, before participatory GIS can be defined, it is first necessary to define GIS. Geographical information systems (GIS) represent a spatial subset of information systems used in urban planning and other land-based or geographical industries. A popular definition of GIS is a 'computing application capable of creating, storing, manipulating, visualising and analysing geographical information' (Goodchild 2000). However, Voss et al. (2004) state that GIS alone cannot solve all the problems in the planning process and that any system needs to include a decision support element. McHaffie (1995) warns against the conflation of new GIS technologies and democracy. It is necessary to consider the purpose and meaning of maps in order to understand how they can be used to support local participation. Whereas maps can be represented as objective artefacts, they can also be viewed as embodiments of power relations (McHaffie 1995).

Traditional GIS tools have been criticised as elitist, relying on experts for their operation and unsuitable for use by the general public (Ghose 2001). These concerns, combined with increases in the availability of Internet technologies, have led to the development of public participation GIS. These have also been called Web-based public participation system (WBPPS) (Peng 2001) and participatory GIS (PGIS) (Nyerges et al. 2002). PPGIS were initially developed in response to the need for greater public participation in planning. However, they have been criticised for their poor usability for non-technical audiences due to their origins as technical GIS and lack of user-centred design (Haklay and Tobón 2003). In addition to this, the public are not the only stakeholders that need to participate in the planning process and the ideal PPGIS should consider a multidimensional stakeholder model (Innes and Booher 2004). Kingston et al. (2000) propose three principles for a Web-based PPGIS to provide equal access to data and information for all sectors of the community; empower the community by providing the necessary data and information which matches the needs of the community who are, or potentially are, participating; and establish and maintain a high degree of trust and transparency within the public realm to give the process legitimacy and accountability.

Kingston explains that a major limitation of their approach to PPGIS development is that a person is required to build each site for each community (Kingston et al. 2000). A characteristic of wicked problems is that each problem is essentially unique and a 'one-shot' opportunity (Rittel and Webber 1973). Therefore, a tool to support Kingston et al's principles would also have to be adaptable to the unique and changing context of different cities and diverse stakeholders. However, the main drawback of current PPGIS is that they are developed in an ad hoc manner for particular contexts and stakeholder groups which limit the potential for communities to adapt the tools themselves to their needs. This is not appropriate in the context of smart cities where standardised and adaptable participatory tools are required to support the planning of citizen-oriented smart cities. Peng (2001) argues that the 'system should be highly scalable, to allow further expansion as more data, models and users evolve' and that 'since the system is targeted towards the general public, a proper system design [is] essential'. Therefore, an approach is needed that includes the perspectives of multiple stakeholders and unique communities. Brabham proposes that a solution to these challenges lies in the software architecture of the information system (IS) (Brabham 2009). The use of information systems to support participation in planning faces challenges which include the digital divide; accessibility; usability; cost; engaging users; boundaries of the geographical community; and the rules for including and excluding participants (Brabham 2009). In the following section, architecture frameworks are investigated to provide a common structured approach to developing participatory IT solutions for smart cities.

3 Architecture Frameworks

Wicked problems are both uncertain and complex (Hartmann 2012). Wicked problems are uncertain because they are ill-defined problems. This is due to unstable requirements and constraints based on ill-defined environmental contexts (Hevner and Chatterjee 2004). The complexity of wicked problems requires inherent flexibility to change the design processes as well as design artefacts (i.e., malleable processes and artefacts) (Hevner and Chatterjee 2004). Cities are made up of loosely tied communities and typically do not have a common vision and strategy. It is therefore problematic that the narrative for smart cities is being set by IT service providers (Söderström et al. 2014) rather than adapting to the requirements of the users of those services. Examples of architecture frameworks in the context of this chapter include Al-Hader et al. (2009) who describe the components of a smart city architecture and McGinley and Nakata (2012) describe an architecture framework for sustainable communities. Additionally, Ebrahim and Irani (2005) propose a four-layer architecture framework for e-government based on a meta-analysis of the adoption barriers from the perspective of government as the service provider. It focuses on creating a 'common way of finding all government information and services' (Ebrahim and Irani 2005). This model suits the corporate narrative of fixing potholes and registering complaints about garbage collection, but it fails to help the communities to develop their own sense of their community online and to discuss the issues they have as community.

However, it can be difficult to develop digital tools to support such contexts as the goals and conflicts of the various stakeholders can be difficult to establish. In businesses, enterprise architecture (EA) frameworks are used to link their needs to their information systems. EA frameworks link the operational goals of an enterprise to, among other things, its information systems. Anthopoulos and Fitsilis (2010) have explored the use of enterprise architecture for digital cities. However, the unique problem of smart cities means that existing enterprise architecture should be expanded to view the city as an enterprise of diverse and loosely connected stakeholders rather than a single enterprise that EA frameworks were designed for. Moreover, Chourabi et al. (2012) call for a framework to support the 'making' of smart cities. Chourabi et al. (2012) identified eight clusters of factors for smart cities (1) management and organisation, (2) technology, (3) governance, (4) policy, (5) people and communities, (6) the economy, (7) built infrastructure and (8) the natural environment. From these, they identified that (1) management and organisation and (5) people and communities are the least considered factors in current approaches to making [planning] smart cities. The technology factor identified by Chourabi et al. (2012) is expanded on Ebrahim and Irani (2005)'s analysis of the technology factors in e-government initiatives, which shows that the information and business artefacts of smart cities would map more easily to EA than a representation of the people and community artefacts.

Therefore, this chapter identifies and presents a 'community architecture' framework that can be used to develop information systems to support diverse community stakeholders to participate in local land use planning. In order to achieve this, it is argued that the framework would need to represent the perspectives and motivations of the diverse stakeholders and their unique communities. There are tools to support public participation in urban planning; however, these have typically been developed in an ad hoc approach. Nyerges and Drew (2001) propose the use of an enterprise architecture framework to address this, but do not test it. In organisations that include an authority to define the strategy, goals and vision of the organisation, it would be possible to use a range of enterprise architecture (EA) frameworks and methods to develop tools to support these strategies. However, among the smart city community stakeholders, typically no single representative exists; therefore, 'off-the-shelf' EA frameworks would be inappropriate in a smart city community context. Innes and Booher (2004) suggest that participatory urban planning requires a multidimensional perspective model. Therefore, there is a requirement for a participatory or 'community' architecture framework, based on an appropriate EA that is flexible enough to incorporate the multiple perspectives, visions and strategies present among community stakeholders. This research will identify and test an appropriate 'community' architecture framework to support the development of tools for the participatory planning of smart cities.

4 Method

This research proposes an architecture framework that is appropriate to all stakeholders of smart cities. Architecture frameworks are most commonly found in business, where enterprise architecture frameworks are typically used to link the needs and systems of business. The aim of this research is to identify an appropriate architecture framework to support the development of tools that support participatory smart city planning. The identified 'community architecture framework' (CAF) will then be tested by applying it to the diverse stakeholder context of participatory urban planning. Given the complexity and length of smart city development, the CAF is tested in this chapter on a simpler case to identify what citizens, planners and developers wish to do with specific spaces in a city that were of interest to them. The case proposed in this chapter is targeted at multiple community stakeholders including the general public, developers and planners. It is essential that the output of the system is useful to all the stakeholders. The developed tool is evaluated in a series of stakeholder interviews. This case study is then used to frame a discussion as to how this approach could be used to support the design of smart cities. The following section describes the development of the CAF.

5 Identifying an Appropriate Architecture Framework

Enterprise architecture (EA) frameworks provide a link between the needs of the organisation and the proposed information system architecture. However, EA frameworks are not designed to consider diverse stakeholder perspectives and therefore in their current form are inappropriate in a smart city context. Therefore, a community architecture (CA) framework that supports diverse stakeholder perspectives needs to be identified (Fig. 1). In order to determine which EA framework is most appropriate to adapt into a CA framework, a definition of community architecture is proposed. This is achieved by adapting a popular EA definition: [EA] *is a representation of the relationship of an organisation's business operations to the processes and data that support them* (Boxer and Garcia 2009). By replacing 'organisation' with 'community', a description for the architecture of the (whole) community is provided. However, the criterion is to develop an architecture definition that supports diverse stakeholder perspectives. Therefore, the proposed definition of community architecture is '*a representation of the relationship of the community stakeholders' perspectives to the processes and data that support them*'.

The Zachman framework (ZF) is 'the de facto standard for classifying the artefacts developed in enterprise architecture' (The Open Group 2006). Sowa and Zachman (1992) describe the framework as a convenient classification scheme or 'periodic table' for information entities. The framework consists of a 6×5 matrix (labelled 'original' in Fig. 2). By filling in its cells, it is possible to specify the

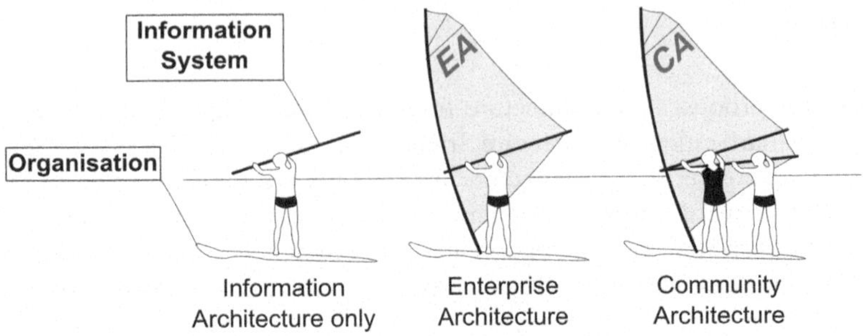

Fig. 1 An illustrative comparison of information, enterprise and community architecture

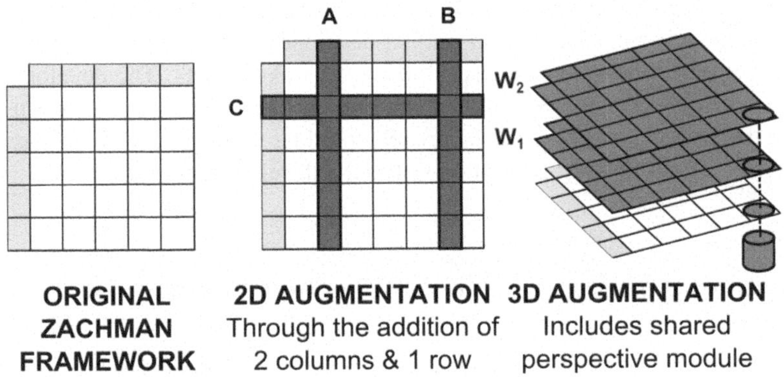

Fig. 2 Zachman framework modification options

artefacts of the architecture and their relationship to each other. The columns are titled (what, how, where, who, when and why), and each one represents a complete architectural model (data, function, network, people, time and motivation). The rows represent perspectives of the system, but not perspectives of the stakeholders. Therefore, in its current form, the Zachman framework does not represent a multidimensional stakeholder model that can respond to the uncertainty, complexity and normativity of wicked problems. An alternative EA is TOGAF (The Open Group Architecture Framework). TOGAF is divided into 10 parts and defines rules for developing robust architecture principles rather than explicitly stating the principles in the framework (Urbaczewski and Mrdalj 2006). Although insufficient in their original form, TOGAF and the Zachman framework both contain elements that are necessary to design information systems in a wicked problem context. Between the two, the intuitive form of the Zachman framework's primary interrogatives would be simpler to implement with non-technical stakeholders than the EA-specific terminology of TOGAF. Therefore, the Zachman framework is appropriate as a basis for the community architecture framework for the diverse stakeholders of smart cities. However, it is clear that the original

Zachman framework needs to be adapted to support diverse stakeholder perspectives. Three approaches to this are discussed in this section which are summarised in Fig. 2.

Gous et al. (2011) provide a comparison of examples of business architectures in interorganisational innovation networks using the Zachman framework. Their approach is modelled on each organisation (or in our case stakeholder), using their own unconnected unmodified Zachman framework (Fig. 2). However, the resulting network's combined pool of goals and motivations is not necessarily strategically aligned. This leads Gous et al. to conclude that the Zachman framework is a 'viable option for structuring architecture information on interorganisational innovation networks'. By extension, it could be argued that if the Zachman framework can work in strategically unaligned contexts it could make a good base to develop a 'community architecture' framework that can work in strategically unaligned multistakeholder networks. However, Gous et al. argue that the geographical distribution of the collaborative network is sustainable only if 'a well-connected and proactive intermediary' is present. Therefore, alternative framework adaptations that can accommodate multidimensional perspectives are explored here.

Boxer and Garcia (2009) argue that enterprise architecture frameworks (EAFs) are designed for organisations with a single source of control. Therefore, in order to use them in distributed multiorganisational contexts with 'decentralised governance', changes would need to be made to standard EAFs in order to make them relevant to the requirements of system in system contexts. Some details of the enterprise architecture would normally be implicit and understood within the context of the enterprise. However, Boxer and Garcia argue that it is important to make the implicit details explicit in order to facilitate communication between the collaborating organisations. To achieve this, they suggest adding a row to the Zachman framework that describes the relationships of the collaborating enterprises (Row C in Fig. 2) as well as to add two columns, the first of which (Column A in Fig. 2) is intended to extend the *data/what* column to include the 'physical reality' of the data that is being modelled upon. The second column (Column B in Fig. 2) is intended to provide a 'for whom' perspective. The Boxer and Garcia approach is relevant to the purpose of this research to support systematic development with non-strategically aligned stakeholders, as is the case with supporting the participatory planning of smart cities. The addition of Row C would be useful in the case of this research, but to include Row A and Row B would lead to redundancy in the resulting framework.

Alternatively, Martin and Robertson (1999) propose using the existing framework as layers in an extended three-dimensional model. This is further explored in Morganwalp and Sage (2003) who explore systems of systems (SoS) representation of the framework that they describe as a '3D enterprise architecture framework'. They approach this by modelling multiple related systems in parallel in the third dimension of a shared 2D framework. They argue that whilst a 2D framework works well at modelling an individual system, it is not appropriate for modelling 'all the systems in an enterprise as a whole' (Morganwalp and Sage 2003,

p. 93). In '3D augmentation' example in Fig. 2, different stakeholder 'workspaces' (Vonk and Ligtenberg 2010) are arranged on top of each other in a non-hierarchical stack of stakeholder workspaces. In addition, a new stakeholder coordination mechanism could be implemented to connect the corresponding cells in the other stakeholder perspectives (workspaces). This would provide an opportunity to pull the common artefacts from each stakeholder's workspace and support multidimensional perspectives for every artefact in the framework.

5.1 Community Architecture Framework (CAF)

The approaches in the previous section show that it is possible to augment the existing 2D Zachman framework either with extra columns and/or rows or alternatively into three dimensions in order to incorporate the multiple perspectives of the stakeholders and have a coordinating intermediary to link the strategically unaligned stakeholders in the network with the data and processes of the system. However, it is not necessary to provide a multidimensional perspective for every artefact by following Innes and Booher (2004)'s multidimensional model too literally. It is clear that the proposed 3D adaption loses some of the simplicity of the initial framework and will lead to redundancy because some cells in the framework do not need to be duplicated. Indeed, it is possible that only the first row of the framework requires a multidimensional stakeholder model. Boxer and Garcia's additional 'collaborative' Row C provides a multidimensional stakeholder perspective.

Therefore, the 2D approach of the additional row of Boxer and Garcia without the additional columns is proposed for the community architecture framework. This approach is described in Table 1 with the additional alteration that instances of the term 'business' from the Zachman framework have been replaced with 'community'. This results in several new terms including 'community process model' and 'community logistics model'. Table 1 presents the proposed community architecture framework and includes an additional row to represent the community architecture collaborative model. This revealed some interesting challenges such as how to capture a 'community vision' for smart cities. The community architecture framework proposed in Table 1 describes each instance of the collaborating perspective of each stakeholder. In this way, the framework builds a picture of a 'unified community' from the presentation of the multiple perspectives, interests and requirements of autonomous stakeholders. In Table 1, the data and other artefacts of the information are visualised by the framework. This makes the link between the resultant framework and the components of the system explicit. In this sense, the framework conforms to community architecture, i.e. *a representation of the relationship of the community stakeholders' perspectives to the processes and data that support them.* In the next section, a system development methodology for the framework is proposed.

Table 1 The proposed community architecture framework

	What	**How**	**Where**	**Who**	**When**	**Why**
Scope *Contextual* planner	List of things important to the community	Community processes	Stakeholder locations	Stakeholders	Community cycles	Community goals
Collaborative model *Collaboration* stakeholder	**S_1: What**	**S_1: How**	**S_1: Where**	**S_1: Who**	**S_1: When**	**S_1: Why**
	S_2: What	**S_2: How**	**S_2: Where**	**S_2: Who**	**S_2: When**	**S_2: Why**
	S_n: What	*S_n: How*	*S_n: Where*	*S_n: Who*	*S_n: When*	*S_n: Why*
Community model *Conceptual* 'owner'	Semantic model	Community process model	Community logistics network	Workflow model	Community events	Community vision
Systems model *Logical* designer	Logical data model	Applications architecture	Distributed systems architecture	Human interface architecture	Process structure	Community rule model
Technology model *Physical* implementer	Physical data model	System design	Technology architecture	Presentation architecture	Control structure	Rule design
Detailed representation *Out-of-context* subcontractor	Data definition	Program	Network architecture	Security architecture	Timing definition	Rule definition
Generic	**Data**	**Function**	**Network**	**Organisation**	**Schedule**	**Strategy**

5.2 Community Architecture Development Methodology

It is not possible to develop tools based on the CAF alone; for this, a community architecture development methodology (CADM) is required. In a design science research methodology (DSRM) to develop information systems (Peffers et al. 2008), this step is referred to as demonstrating 'the use of the artefact [from the design and development step] to solve one or more instances of the problem'.

The following sections are organised into the steps of the CADM described in Fig. 3, covering the following steps: (1) a mock-up is developed and evaluated based on the concepts identified for the tool; (2) a series of requirements and features are proposed based on 1; (3) the community, system and technical architectures for the features are proposed; (4) the features are implemented and evaluated; (5) the results are analysed; and (6) the results are communicated.

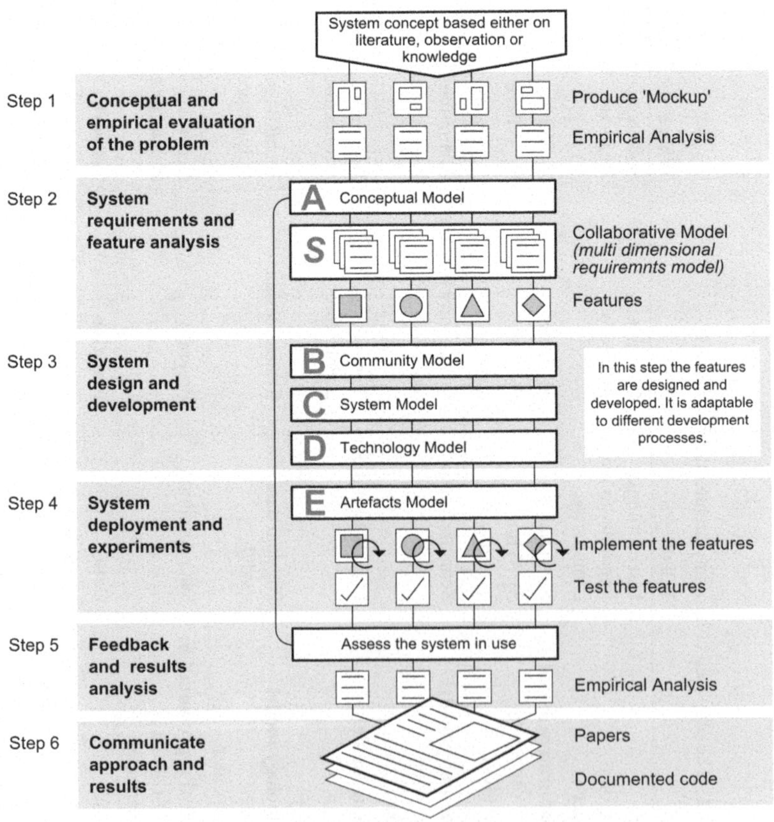

Fig. 3 The community architecture development methodology (CADM)

6 PartiCity Development

The community architecture framework (CAF) and community architecture development methodology (CADM) proposed in the previous section are used here to develop an online tool to support multistakeholder participation in planning. The tool is called PartiCity, a portmanteau of participation and city. Although the tool is tested in one county in the UK in the following section, it is intended to be applicable to any geographical community. Following step 1 of the CADM, the concepts of the tool are identified, and for this case study, they were ROLE, USER, COMMUNITY, SITE, CHALLENGE, GOAL and IDEA. Whilst not seemingly smart city-specific concepts, they are designed to generically support a variety of urban planning issues. Based on the conceptual analysis of the tool, an entity–relationship diagram (ERD) was produced (Fig. 4). Based on the entity–relationship diagram, a mock-up was produced which is shown in Fig. 5.

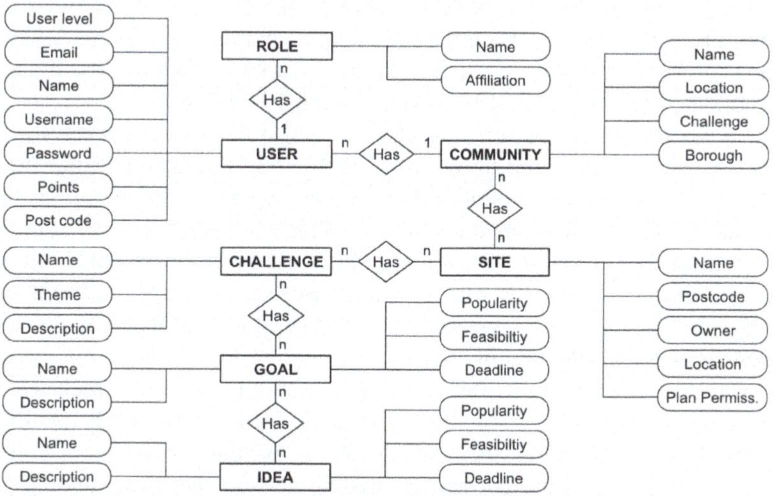

Fig. 4 Entity–relationship diagram for the PartiCity mock-up

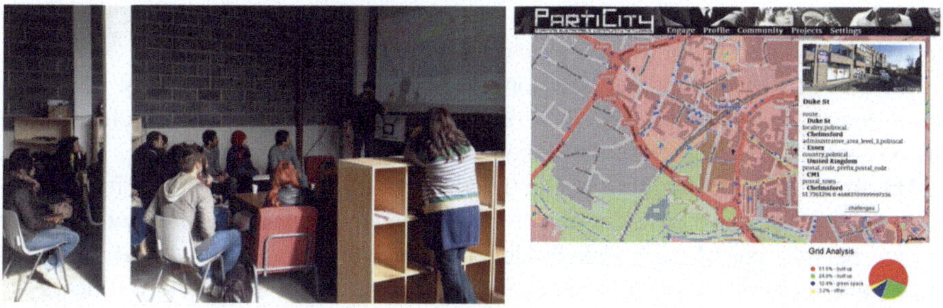

Fig. 5 Empirical analysis of the mock-up based on the concepts derived from the ERD (Fig. 4)

Table 2 Initial state of the semantic table

Term	Public	Planner	Developer
Community	Locals	Community	Community
Development	Idea	Idea	Idea
Sites	Place	Site	Plot

The mock-up focussed on capturing the problem and challenges faced by the whole community, ultimately intending to accumulate a composite community vision. It enabled the user to define their community from anywhere in the UK, but offered a limited and fixed set of challenges focused on the needs of the citizen stakeholder. The mock-up tool was tested in a focus group of 20 participants at a case study in Chelmsford, a town in eastern England. One of the participants requested that they wanted the ability to define the challenges for their community. Another participant suggested that the selection of different roles in the process should trigger different responsibilities. Based on the empirical analysis of this mock-up, the collaborative stakeholder model was populated in step 2 which enabled the stakeholders to be modelled separately. Table 2 describes the initial state of the semantic table (collaborative model column 1 in Table 1).

This shows that there are some similarities and differences between the stakeholders, i.e. they use different terms for COMMUNITY and SITES but the same term for IDEA, which were identified as concepts in the entity–relationship diagram as in Fig. 4. This demonstrates how the semantic table could be used by the tool to look up the relevant term for a specific concept based on the ROLE of the USER. The feedback from the first iteration was used to adapt the framework to enable the development of a system that could support multiple stakeholder perspectives through the selection of a role.

By using the framework, it was possible to view the changes which were then visible in the framework and in the system. A second case study at Sheffield University, UK, revealed that the 'design' of the system for the second iteration was 'untrustworthy' and the participants recommended that the branding should be made more neutral and 'trustworthy'. Step 2 also generated the requirements for the tool, which were as follows: (1) the tool should support unique community contexts; (2) the process should be transparent and trustworthy; (3) the tool should be accessible, immediate and easy to use; (4) the tool should support the user's perspective; (5) the output of the process should be useful to the users;

(6) The process should support conflict whilst aiming for consensus. The sixth requirement was supported by the collaborative model (Table 3) which describe the possible goals for each of the stakeholders (Fig. 6).

Based on an understanding of the diverse goals of the stakeholders, it was possible to develop a use case diagram for idea management using the tool from the unique perspective of each stakeholder. In step 3, the community model, system model and technology model are developed. The software architecture follows the client–server model as shown in Fig. 7.

Table 3 The goals of each of the stakeholders

Stakeholder	Possible goals
Public	Object or support proposed ideas Inspire development and interest in areas
Planner	Consult the public on (an existing set of ideas) Focus developer attention (on an area of interest to the planners)
Developer	Consult the public (in order to get planning permission) Make a return on their investment (get planning permission) Make a return on their investment (develop the land/place/building)

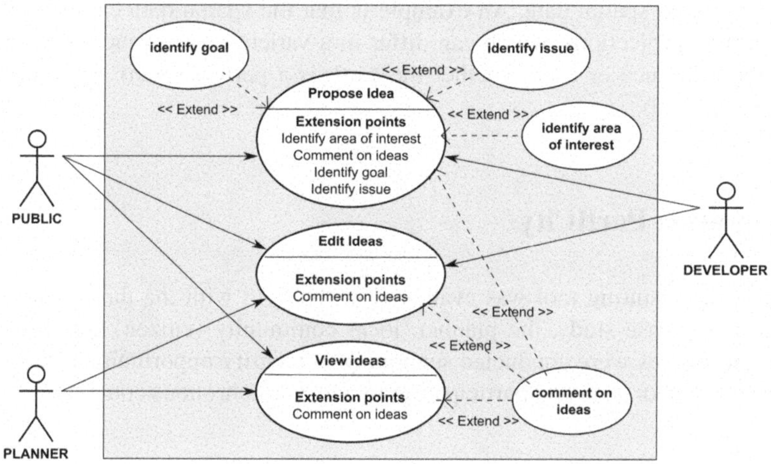

Fig. 6 Use case diagram for PartiCity

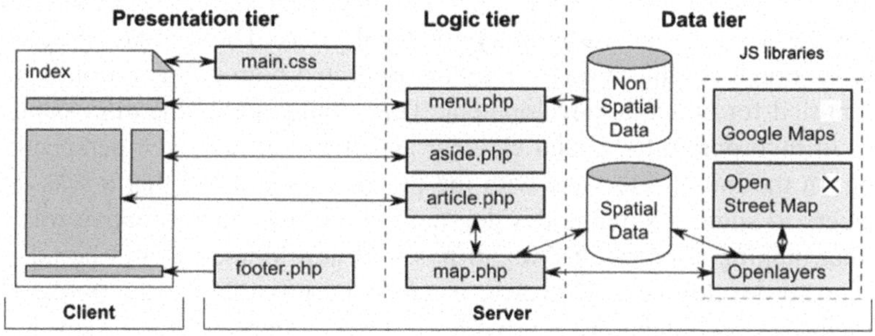

Fig. 7 PartiCity 3-tier software architecture

In order to reduce the time to load, PartiCity utilises a single-page architecture that uses AJAX to update the side bar and article components of the new components as needed without having to reinitialise the map or menu bar for instance. The prototype feature is mainly coded in PHP using HTML5 standards. JavaScript (JS) libraries provide the client-side requests to external data such as the map tiles. Batty et al. (2010) caution that whilst Web 2.0 and 'mash-up' technologies present new opportunities for users to develop their own map content, the skills required to do this mean there is little potential for non-experts to be able to use the tools to design their own communities. An example of the challenge of developing PPGIS is in mastering heterogeneous data sources from disparate sources. This is more challenging with spatial data. An example is that the spatial data can be stored in different map projections, which can differ in a variety of ways including the origin of the reference or even whether the earth is a perfect sphere or an ellipsoid (Batty et al. 2010).

7 Evaluating PartiCity

In step 4, the resulting tool was evaluated in the field with the three main stakeholders of the case study, the planner, local community (citizen) and developer. These case studies were conducted separately to identify opportunities for the tool to support collaboration and participation in a non-synchronous context.

7.1 The Planner's Perspective

From the perspective of the planner that was interviewed, they were interested in capturing the ideas of the community for local areas. These areas represented a potential resource conflict because they incorporated both conservation zones and sites identified for potential development. This would enable the discussion about the sites to incorporate the social element and therefore all three perspectives on planning. In the initial interview with the planner, they stated that it was difficult for planners to suggest suitable development due to their authoritative role which meant that making 'suggestions' about possible development could be taken as a permission to development. However, it would be possible for the public to propose suitable development. This provides a good example of how the role of the public, planners and developers needs to be supported differently by the tool.

Figure 8 shows the input screen for the planner. The planner screen has no function to input specific sites. Instead, they try to capture challenges at a community rather than site or place level. Therefore, it was more appropriate for the planner to identify the challenges the location faced. The tool to capture the input of the planners is represented in Fig. 8.

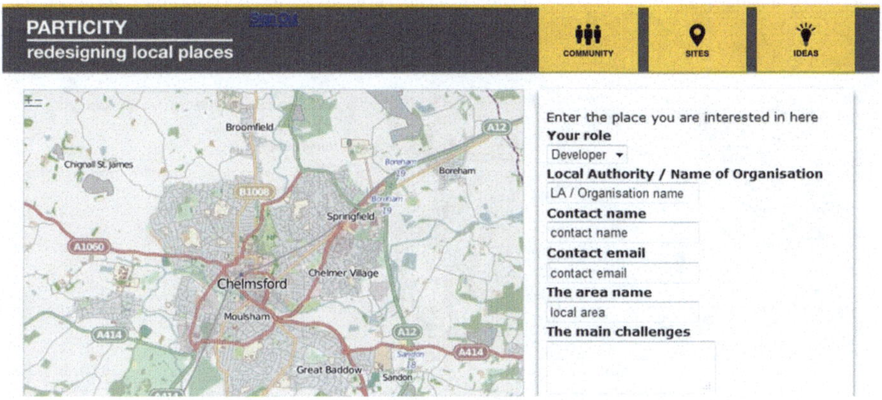

Fig. 8 Page to enable planners and developers to input the area they are interested in

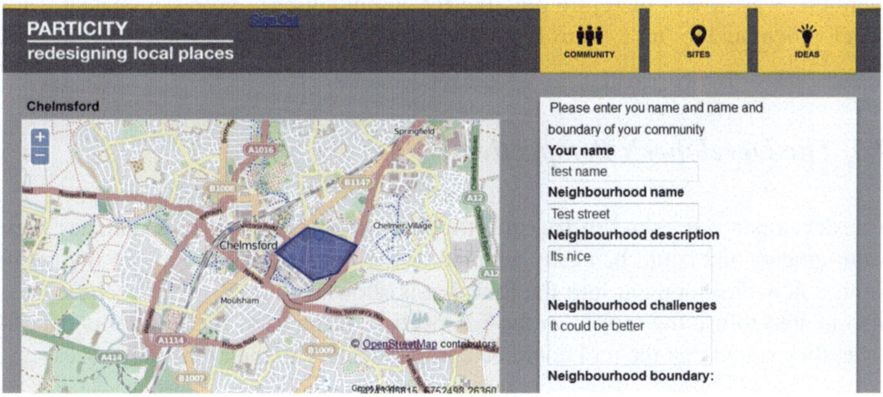

Fig. 9 The public page to identify a user's community

7.2 The Public's Perspective

A prototype of the standardised community capture feature was developed, and the public were invited to try it out and then interviewed. The first stage of the process involved the user inputting information about their community (Fig. 9). Users were asked to complete the information about their community which included its challenges within its spatial boundary, its name and a description of it.

Following this, participants were shown the next page in the process which enabled them to input specific sites and places in their communities whereby specific issues of their communities could be addressed (Fig. 10). In both cases, the areas of interest are highlighted in blue.

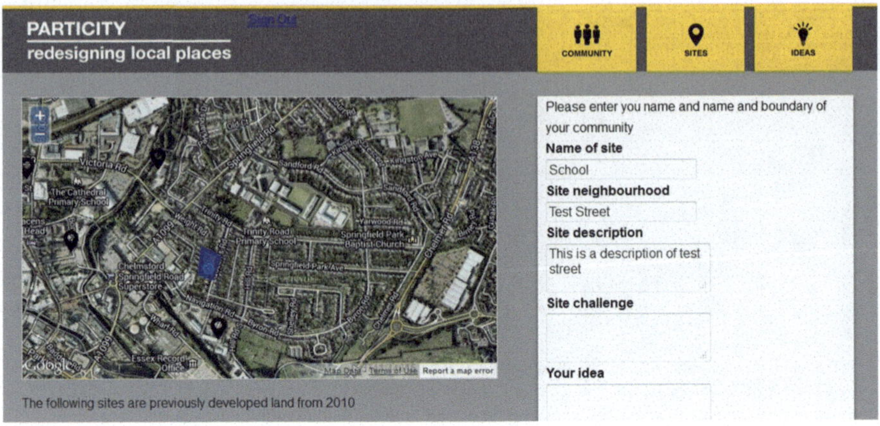

Fig. 10 The public page to identify a user's idea for a site

In this way, it was therefore possible for communities to propose local sites for development and to 'take control' of the discussion regarding that site.

7.3 The Developer's Perspective

The developer was asked to trial the tool and to assess whether it matches their requirements and could be easily adapted to their processes. They were invited to input a new development into the tool that is shown in Fig. 8. The developer was also invited to use the tool from the perspective of the user. Based on this experience, they stated that the tool could be complementary to their existing processes.

8 Discussion

PPGIS has heralded the potential for a new Web technology-enabled era of participation in urban planning. Brabham (2009) goes further and argues for 'crowdsourcing' public participation in urban planning. Brabham calls for 'an embrace of technological solutions' in order to harness the crowdsourcing potential of the Web to plan the urban environment. Consensus building among stakeholders requires 'broad access to information', and one of the biggest issues in planning is information (Hanna 2000). This issue can be summarised as who controls the information and how trustworthy is it (Innes and Booher 2004). It is therefore important to build tools to support participation in planning that represents the diverse perspectives in cities. In step 5, the system was tested by community, planner and developer representatives. This section forms step 6 wherein the results of the case study are presented, and in this case, the implications for the development to support participation in smart city planning is discussed. Three different interfaces were developed in this research to represent the different perspectives, but these were built upon the same computational back end (Fig. 7). The development of the 'PartiCity' tool and its testing in the case studies above show that it was possible for diverse stakeholders to record their interests into a common participatory tool. This approach therefore has demonstrated that it facilitates the development of a tool to support participation in the planning of smart cities. The initial mock-up questioned the public on specific ideas that the planners or developers already had in the pipeline or on a 'wish list'. However, the most valuable point to come out of the case studies was the idea of a public wish list of projects that people of self-identified geographical communities could put forward their own ideas for the future of their city. In the case of the step 4 mock-up (Fig. 10), the users were allowed to input any idea that could be linked to a spatial boundary. However, in future work in smart cities, a themed approach that allows users to both comment on existing options and make their own suggestions for the future would be appropriate. In parallel with this study, we are exploring the application of the community architecture framework in built environment architecture and hope to ultimately bring the social, digital and physical layers of smart city built environments together through a participatory process using the community architecture framework.

9 Conclusion

This chapter presented the development of a 'community architecture' framework (Table 1), which was used to develop a tool to support developers, planners and communities to participate in the planning of their cities by proposing new ideas for their areas of interest. This chapter demonstrated that by using the community architecture framework, it was possible to retain the unique goals and perspectives of the stakeholders, thereby allowing them to find common ground without having to enforce standardised options for them to consider. This is in contrast to traditional enterprise architecture frameworks which seek to represent the structured goals and perspective of a single organisation. This is important as it offers opportunities for innovation in the planning of smart cities which would be challenging in a more rigid model. Based on this work, this chapter proposes that the community architecture framework could be used to develop a system to support the participatory design of smart cities. The PartiCity tool presented here would have similar requirements to the smart city design tool. The community architecture framework does seem to be appropriate to the challenges inherent in supporting participation in smart city design. However, a complete community architecture development methodology (CADM) cycle would need to be undertaken on the development of a smart city design tool on a live smart city project in order to test this further. Due to the prohibiting scale of such an undertaking, it was appropriate to test the approach in the prototype form presented here first before embarking on a 'live' project of the scale and importance of smart cities. This work demonstrates that there is no excuse for the goals and requirements of diverse city stakeholders not to be included in the development of tools to support the design of smart cities. This could lead to the realisation of an exciting and progressive alternative narrative for smart cities.

Acknowledgements Chelmsford City Council Planning Department, Chelmsford Ideas Hub and Vera Hale at Sheffield University for supporting the focus group. This research was supported by the EPSRC in the UK.

References

Al-Hader, M., Rodzi, A., Sharif, A. R., & Ahmad, N. (2009). Smart city components architecture. In *IEEE International Conference on Computational Intelligence, Modelling and Simulation* (pp 93–97).

Anthopoulos, L., & Fitsilis, P. (2010). From digital to ubiquitous cities: Defining a common architecture for urban development. In *Sixth International Conference on Intelligent Environments* (pp. 301–306).

Arnstein, S. R. (1969). A ladder of citizen participation. *Journal of American Institute of Planners, 35*, 216–224. doi:10.1080/01944366908977225.

Australian Public Services Commission. (2007). *Tackling wicked problems: A public policy perpective.*

Batty, M., Hudson-Smith, A., Milton, R., & Crooks, A. (2010). Map mashups, Web 2.0 and the GIS revolution. *Annals of GIS*, *16*(1), 1–13. http://doi.org/10.1080/19475681003700831.

Boxer, P. J., & Garcia, S. (2009). Enterprise architecture for complex system-of-systems contexts. In *3rd Annual IEEE Systems Conference* (pp. 253–256). doi:10.1109/SYSTEMS.2009.4815807.

Brabham, D. C. (2009). Crowdsourcing the public participation process for planning projects. *Planning Theory, 8*, 242–262. doi:10.1177/1473095209104824.

Campbell, S. (1996). Green cities, growing cities, just cities?: Urban planning and the contradictions of sustainable development. *Journal of the American Planning Association, 62*, 296–312. doi:10.1080/01944369608975696.

Chourabi, H., Nam, T., & Walker, S., et al. (2012). Understanding smart cities: An integrative framework. In *45th Hawaii International Conference on System Science* (pp. 2289–2297).

De Cindio, F., De Marco, A., & Ripamonti, L. A. (2007). Enriching community networks by supporting deliberation. In *Communities and technol*ogies (pp 395–417). London: Springer.

Ebrahim, Z., & Irani, Z. (2005). E-Government adoption: Architecture and barriers. *Business Process Management Journal,*. doi:10.1108/14637150510619902.

Ghose, R. (2001). Use of information technology for community empowerment: Transforming geographic information systems into community information systems. *Transactions in GIS, 5*, 141–163. doi:10.1111/1467-9671.00073.

Giffinger, R., Fertner, C., & Kramar, H., et al. (2007). Smart cities ranking of European medium-sized cities. doi:10.1016/S0264-2751(98)00050-X.

Goodchild, M. F. (2000). Spatial analysts and GIS practitioners the current status of GIS and spatial analysis. *Journal of Geographical Systems, 5*, 5–10.

Gous, H., Gard, J., & Baltes, G., et al. (2011). Business architecture for inter-organisational innovation networks : A case study comparison from South Africa and Germany. In K. D. Thoben, V. Stich & A. Imtiaz A (Eds.), *Proceedings of 17th International Conference on Concurrent Enterprising (ICE 2011)* (pp. 1–11). Aachen, Germany: IEEE.

Haklay, M., & Tobón, C. (2003). Usability evaluation and PPGIS: Towards a user-centred design approach. *International Journal of Geographical Information Science, 17*, 577–592. doi:10.1080/1365881031000114107.

Hanna, K. S. (2000). The Paradox of Participation and the Hidden Role of Information: A Case Study. *Journal of the American Planning Association, 66*(4), 398–410. http://doi.org/10.1080/01944360008976123.

Hartmann, T. (2012). Wicked problems and clumsy solutions: Planning as expectation management. *Planning Theory, 11*, 242–256. doi:10.1177/1473095212440427.

Hevner, A., & Chatterjee, S. (2004). Design science research in information systems. In A. Hevner & S. Chatterjee (Eds.), *Design research information systems integrated* (pp. 9–23). US: Springer.

Hollands, R. G. (2008). Will the real smart city please stand up? *City, 12*, 303–320. doi:10.1080/13604810802479126.

Innes, J. E., & Booher, D. E. (2004). Reframing public participation: Strategies for the 21st century. *Planning Theory and Practice, 5*, 419–436. doi:10.1080/1464935042000293170.

Kingston, R., Carver, S., Evans, A., & Turton, I. (2000). Web-based public participation geographical information systems: An aid to local environmental decision-making. *Computers, Environment and Urban Systems, 24*, 109–125. doi:10.1016/S0198-9715(99)00049-6.

Maquil, V., Psik, T., Wagner, I., & Wagner, M. (2007). Expressive interactions—supporting collaboration in urban design. In *Proceedings of International ACM Conference on Supporting Group Work—Gr '07 69*. doi:10.1145/1316624.1316635.

Martin, R., & Robertson, E. L. (1999). Formalization of multi-level Zachman frameworks. *Computer Science Department, Technical Report.*

McGinley, T., & Nakata, K. (2012). A participatory design approach to the wicked problem of designing sustainable communities. In *TSBE EngD Conference.*

McHaffie, P. H. (1995). Manufacturing metaphors: Public cartography, the market, and democracy. *Ground Truth*, 113–129.

Morganwalp, J., & Sage, A. P. (2003). A system of systems focused enterprise architecture framework and an associated architecture development process. *Information, Knowledge, sytems Manag, 3*, 87–105.

Nyerges, T., & Drew, C. (2001). *Considerations for an enhanced decision mapping system : An information system architecture approach*. Citeseer: Princeton.

Nyerges, T., Jankowski, P., & Drew, C. (2002). Data-gathering strategies for social-behavioural research about participatory geographical information system use. *International Journal of Geographical Information Science, 16*, 1–22. doi:10.1080/13658810110075987.

Peffers, K. E. N., Tuunanen, T., & Rothenberger, M. A. (2008). A design science research methodology for information systems research. *Journal of Management Information Systems, 24*, 45–77. doi:10.2753/MIS0742-1222240302.

Peng, Z. R. R. (2001). Internet GIS for public participation. *Environment and Planning B Plan Design, 28*, 889–905. doi:10.1068/b2750t.

Rinner, C., & Bird, M. (2009). Evaluating community engagement through argumentation maps—a public participation GIS case study. *Environment Planning B Plan Design, 36*, 588–601. doi:10.1068/b34084.

Rittel, H. W. J., & Webber, M. M. (1973). Dilemmas in a general theory of planning. *Policy Sciences, 4*, 155–169. doi:10.1007/BF01405730.

Sieber, R. (2006). Public participation geographic information systems: A literature review and framework. *Annals of the Association of American Geographers, 96*, 491–507. doi:10.1111/j.1467-8306.2006.00702.x.

Söderström, O., Paasche, T., & Klauser, F. (2014). Smart cities as corporate storytelling. *City, 18*, 307–320. doi:10.1080/13604813.2014.906716.

Sowa, J. F. F., & Zachman, J. A. (1992). Extending and formalizing the framework for information systems architecture. *IBM Systems Journals, 31*, 590–616. doi:10.1147/sj.313.0590.

The Open Group. (2006). *ADM and the Zachman framework*. http://pubs.opengroup.org/architecture/togaf8-doc/arch/chap39.html.

Urbaczewski, L., & Mrdalj, S. (2006). A comparison of enterprise architecture frameworks. *Issues in Information Systems, 7,* 18–23.

Vonk, G., & Ligtenberg, A. (2010). Socio-technical PSS development to improve functionality and usability—sketch planning using a maptable. *Landscape and Urban Planning, 94,* 166–174. doi:10.1016/j.landurbplan.2009.10.001.

Voss, A., Denisovich, I., Gatalsky, P., et al. (2004). Evolution of a participatory GIS. *Computers, Environment and Urban Systems, 28,* 635–651. doi:10.1016/j.compenvurbsys.2003.12.003.

Weber, E. P., & Khademian, A. M. (2008). Wicked problems, knowledge challenges, and collaborative capacity builders in network settings. *Public Administration Review, 68,* 334–349. doi:10.1111/j.1540-6210.2007.00866.x.

Epilogue

Urbanizing Technology

Saskia Sassen
Columbia University, USA
sjs2@columbia.edu

Across time, cities have complicated the straightforward implementation of technologies. The mix of urban materialities and people's cultures in the city is not quite predictable and hence can unsettle or disrupt the best designs. This holds at many levels—from advanced transportation systems to "intelligent systems" installed in buildings, to name just a few. Therewith, the city also is a lens that allows us to understand the diverse interactions between users (whether systems, organizations, or people) and the design and implementation of the technologies used in cities.

The DNA of the city is more akin to open-source technology. An approach that takes this factor into account would enable interactions between the technology and the user beyond those already pre-programmed within these systems. This would strengthen an understanding of the city as a combination of incompleteness and complexity: It is this mix that has enabled cities to outlive enterprises, kingdoms, nation-states, and, yes, Cisco Systems. The notion of the "perpetual beta" also comes to mind. These are all rather closed formal systems, which has made them rigid and more susceptible to obsolescence. One implication is that the current practice of installing more and more closed, centrally controlled intelligent systems in cities puts those cities themselves at risk of becoming obsolete when the technologies become obsolete.

The city puts technologies to a test. It is one window into understanding successful technological innovations for urban systems and urban life. Powerful actors can remake cities in their image. But cities talk back. They do not take it sitting down. Sometimes, it may take decades, and sometimes, it is immediate. The chapters in the *Urban Interfaces* section of this book provide such examples. We can think of the multiple ways in which the city talks back as a type of open-source urbanism: The city as partly made through a myriad of interventions and little changes from the ground up. There are diverse versions of this: DIY

M. Foth et al. (eds.), *Citizen's Right to the Digital City*,
DOI 10.1007/978-981-287-919-6

urbanism, tactical urbanism, urban guerilla tactics, urban acupuncture, and urban prototyping. Each of these multiple small interventions may not look like much, but together they give added meaning to the notion of the incompleteness of cities and that this incompleteness gives cities their long lives, thereby outlasting other more powerful entities.

Beyond the imagery of open-source urbanisms, can we strengthen this positive scenario of the city's incompleteness by actually deploying open-source technologies in a variety of urban contexts? Can we urbanize open-source technology? As a technical practice, open source has not quite been about cities, but about technology itself. Yet it resonates with what cities are at ground level, where its users are. The park is made not only with the hardware of trees and ponds, but also with the software of people's practices.

Ours is a time when cities have become a key space for large-scale adoption of new technologies. Major advances in building and other technologies have left a massive imprint on urban space. This is perhaps most visible in the sharp increases in density and networked systems that the new technologies have made possible. Many of these technologies are not necessarily developed with the urban environment in mind, and in many cases, this does not matter. But in some cases, it does matter and then the challenge becomes how to urbanize these technologies. What might it mean to urbanize technology? One critical feature that both requires is an outcome of citizens exercising their right to the city, including the digital city.

In many ways, cities tend to urbanize technologies, since it is still not quite feasible to simply plop down a new technology in urban space. It requires modifications and mediations. But a closer look suggests that these modifications have to do with overcoming rigidities and risks, especially risks catalogued by insurance companies. This mode is then only vaguely one of urbanizing the technology. It will take more than the logic of safety.

One key dimension of having a genuinely smart city is open sourcing a whole range of systems. I am particularly interested in open-sourcing neighborhoods to bring the specific knowledge about a city that is rooted in knowing one's neighborhood into the larger task of smarting up the city. A city needs to mobilize its own local knowledge base to understand how to use technical developments best for its conditions. Most cities in the world are poor, as is nicely illustrated by the case studies presented in the *Citizen Action* section of this book. Hence, they must find ways of smarting up their cities—on the environmental vector and on the interconnectivity vector—that are reasonably priced and can mobilize the citizenry into participating in these efforts. One risk is that the large corporations that dominate the smart cities business are keen on selling their systems and hence ready to persuade politicians that they should go for these expensive and often not very useful options.

Getting to the heart of these issues requires not only understanding the particular features of cities, but also "seeing" the issue as if one were a city—in other words, juggling the diversity of elements that constitute urban space, which inevitably means a multi-perspective approach. The participatory and collaborative examples in the *City Making* section of this book discuss such approaches.

Elsewhere, I have theorized this as the city having speech.[1] Speech is a foundational element in theories about democracy and the politics. As a concept, it has seen both expansions and contractions of its meaning. But it has not yet been expanded to include the concept that the city might have speech.

The city is a generous partner in this work: It offers a lens into larger realities, many non-urban, that now have an urban moment in their trajectories. This type of analysis keeps us from simply being technologists. It helps us factor in the friction and obstacles that even the most advanced technologies or engineering confronts in urban environments.

[1]Does the City Have Speech?, *Urban Challenges*: Essay, 2013.

Index

M. Foth et al. (eds.), *Citizen's Right to the Digital City*,
DOI 10.1007/978-981-287-919-6

The manufacturer's authorised representative in the EU is Springer Nature Customer Service Centre GmbH, Europaplatz 3, 69115 Heidelberg, Germany. If you have any concerns regarding our products, please contact ProductSafety@springernature.com

Printed and bound by CPI Group (UK) Ltd, Croydon, CR0 4YY
25/09/2026
02222891-0005